# Essential Fluid Dynamics for Scientists

# Essential Fluid Dynamics for Scientists

**Jonathan Braithwaite**

Morgan & Claypool Publishers

ISBN    978-1-6817-4597-8 (ebook)
ISBN    978-1-6817-4596-1 (print)
ISBN    978-1-6817-4599-2 (mobi)

DOI    10.1088/978-1-6817-4597-8

Version: 20180101

IOP Concise Physics
ISSN 2053-2571 (online)
ISSN 2054-7307 (print)

A Morgan & Claypool publication as part of IOP Concise Physics
Published by Morgan & Claypool Publishers, 1210 Fifth Avenue, Suite 250, San Rafael, CA, 94901, USA

IOP Publishing, Temple Circus, Temple Way, Bristol BS1 6HG, UK

*This work is dedicated to Dani,
who put up with me while I was writing it.*

# Contents

# Preface

This book grew out of a course in hydrodynamics taught in Bonn to Master's students. Although the course was designed for students who have already completed an undergraduate degree in physics, only a little prior knowledge is assumed, namely that the student is familiar with very basic thermodynamics and vector calculus. This course is accessible to, and useful for, students in any branch of physics. However, there is some bias towards applications in astrophysics and geophysics, as opposed to applications in engineering. In practice this means there is a certain focus on phenomena important in these fields, such as the types of waves that might be found in stars or in the ocean. In contrast, topics such as boundary layers, pipe flow and aerodynamics, are mentioned only relatively briefly. In addition, many of the principles are illustrated using examples from everyday experience, as in this way the student can develop an intuitive understanding which can then be applied in other contexts. For instance, reference is made to the hydraulic shock formed as water from a tap spreads out across the surface of a wash basin—making a connection to the phenomenon of astrophysical shocks. Phenomena in atmospheric physics are also used as a bridge between terrestrial intuition and the astrophysical context. In fact, since atmospheric fluid dynamics has a longer history than astrophysical fluids, astrophysicists are well advised to learn from this neighbouring field to avoid reinventing the wheel.

*How to read this book*: each chapter is relatively independent and so it shouldn't be necessary to read them in order. Various concepts from the first two chapters are used in the rest of the book, but the student who already has some familiarity with the subject can skip directly to the later chapters and refer back to these two chapters if the need arises. Also, the reader is advised to look at the exercises even if not actually wanting to work through them, as they introduce some useful concepts not contained in the main text.

*Magnetohydrodynamics*, an extension of hydrodynamics to electrically conducting fluids, is introduced in the last two chapters. There is an emphasis here on astrophysical applications—here we cannot draw on terrestrial intuition and must rely on theory. These chapters are 'stand alone' in that they can be read without the rest of the book.

Jonathan Braithwaite
*Munich, August 2017*

# Author biography

## Jonathan Braithwaite

While studying physics at Wadham College, Oxford, Jonathan Braithwaite took up gliding. Wanting to get even higher and thinking the stars might have something to teach us, he entered astrophysics. He obtained his PhD at the Max-Planck-Institut für Astrophysik in Munich in cooperation with the Universiteit van Amsterdam. The stars are theoretically the same everywhere so he was able to follow this up with a stint at the Canadian Institute for Theoretical Astrophysics in Toronto before moving back to Germany to work at the university in Bonn, where he taught the course on hydrodynamics that led to this book. His research has concentrated on hydrodynamics and magnetohydrodynamics in a number of astronomical contexts including stars of various kinds as well as the interstellar and intergalactic media, with a particular focus on the magnetic fields of upper-main-sequence stars and neutron stars. In 2015 he left full-time research to devote his time to various environmental and international development projects as a matter of the heart. His website can be found at http://jonbraithwaite.com.

IOP Concise Physics

# Essential Fluid Dynamics for Scientists

**Jonathan Braithwaite**

# Chapter 1

## Introduction

We first examine what is meant by a fluid before deriving the equations of motion.

### 1.1 The fluid approximation

The ancient Greeks amongst others debated over 2000 years ago whether matter is made from discrete particles or is a continuum, divisible ad infinitum. This question was not properly resolved until well into the 19th century (Brownian motion, etc), by which time useful theories of thermodynamics had already been developed, driven largely by the need to build more efficient steam engines. Therefore, it is not necessary to think about particles in order to understand thermodynamics; it is just necessary to accept a small number of experimentally-supported axioms (the laws of thermodynamics) and the rest of classical thermodynamics follows[1]. Later, when statistical mechanics was developed, it became possible to understand where the laws of thermodynamics come from, in terms of more fundamental physics. However, for practical purposes this is unnecessary and complicates matters.

The same is true of hydrodynamics, the study of fluid flow, which was also developed prior to the conclusion of the atom versus continuum debate. In many situations it is sufficient to treat a fluid as a continuous substance. Now that we know that fluids are made of particles, we can explain some fluid phenomena in terms of more fundamental physics, for instance we can predict the viscosity of a gas (a macroscopic quantity) by consideration of particles, mean-free paths and so on. However, in this book, we shall cover *classical* hydrodynamics, meaning without consideration for the particle nature of matter, making only occasional reference to particles. Before deriving the equations of hydrodynamics, it is useful to look at the *fluid approximation*, how it is built up, and its limitations, so that we know not to try to use hydrodynamics where it does not apply.

---

[1] This book does assume some prior knowledge of thermodynamics; a selection of relevant results can be found in appendix A.5.

doi:10.1088/978-1-6817-4597-8ch1 © Morgan & Claypool Publishers 2017

To describe a body of fluid precisely, in principle the first step would be to write down the Schrödinger equation. Fortunately it turns out that quantum effects can normally be ignored so that we can consider just the position and velocity of each particle. The next approximation is to describe the particles statistically rather than individually, in terms of a distribution function $f$:

$$\delta N(t) = f(\mathbf{r}, \mathbf{u}, t)\delta \mathbf{r}^3\, \delta \mathbf{u}^3, \tag{1.1}$$

where $\delta N$ is the number of particles in a small volume in position/velocity space at time $t$; $\mathbf{r}$ is the space coordinate (a vector with as many components as the space has dimensions) and $\mathbf{u}$ is the velocity. We can then express the conservation laws (mass, momentum, energy) in terms of a Boltzmann equation.

In this system, a small volume in physical space (i.e. $\delta \mathbf{r}^3 = \delta x\, \delta y\, \delta z$) can contain particles with completely different velocities, and this velocity distribution is described in detail at every location in space. In contrast to this, the fluid approximation describes the system in the following way. First we integrate $f(\mathbf{r}, \mathbf{u}, t)$ over all velocity space to obtain a space density $n(\mathbf{r}, t)$, and then we introduce a mean velocity $\bar{\mathbf{u}} = \bar{\mathbf{u}}(\mathbf{r}, t)$, the mean velocity of the particles[2] at position $\mathbf{r}$. This is arrived at by integrating $f(\mathbf{r}, \mathbf{u}, t)\mathbf{u}$ over velocity space and dividing by $n(\mathbf{r}, t)$. (Hereafter the bar on the mean velocity is dropped.) Obviously in doing this we have lost all information about the spread of particle velocities about the mean. However, we can make up for this by noting that if we have local thermodynamic equilibrium (LTE), there is only one degree of freedom in the spread of velocities which we characterise with temperature $T = T(\mathbf{r}, t)$.

In assuming LTE, we are assuming, in effect, that the particles in some small volume are able come into equilibrium with each other via collisions, rather than wandering larger distances before this has been achieved; only in this way can temperature be defined locally. For this condition to hold, it is necessary that the mean free path of the particles is significantly less than any other length scales of interest to us. For instance, in the Earth's atmosphere the mean free path is of order $10^{-5}$ cm while the smallest length scales of interest to us in weather forecasting are perhaps 100 m, so that we may safely treat air as a fluid. In some contexts the fluid approximation is not applicable, for instance in the solar wind where the mean free path of protons is $10^{15}$ cm $\approx 20$ AU (an astronomical unit is the mean distance between the Sun and the Earth). This example brings us to another point: in the fluid approximation we are assuming that all particle species making up a fluid are in LTE amongst themselves and with each other. In other words, all species have the same mean velocity $\mathbf{u}$ and temperature $T$ at any point in space and time. In the solar wind, the electrons have a significantly shorter mean free path and may come into thermal equilibrium with each other while the protons can still be considered collisionless. A proper study of these phenomena is outside the scope of this book; the interested student should consider consulting a book on plasma physics.

---

[2] If the particles do not have uniform mass, we take a mass-weighted mean. This ensures that the resulting equations respect conservation of momentum.

Finally, it is worth noting that the equations of hydrodynamics that we derive using the fluid approximation can sometimes predict situations which violate the applicability of the approximation. A good example of this is shocks (chapter 6)—the fluid equations predict in some circumstances the appearance of discontinuities in the fluid quantities such as **u** and $T$. The relevant length scale in the fluid goes to zero, which is clearly less than the mean free path and violates the fluid approximation! Fortunately there is a way out of this apparently unpleasant predicament without completely abandoning the fluid picture. (In reality, the discontinuity has a thickness roughly equal to the mean free path.)

## 1.2 The hydrodynamic equations

In this section the equations of hydrodynamics are derived[3].

We know from thermodynamics that the state of a fluid can be described in terms of a number of 'functions of state', which in a simple fluid is two, for instance pressure and temperature; all other variables, for instance density or entropy, can be found from the equation of state. In the following we use a simple fluid, but the equations can easily be generalised to include more complex fluids such as a fluid in which the mean molecular weight is not fixed, which one encounters sometimes in astrophysics, or the salinity in an ocean or water vapour concentration in the atmosphere, for example. Note that these quantities are called *intensive variables* as they can be defined and measured at any particular point in space, as opposed to *extensive variables* such as volume or mass which are properties of a whole system. In addition to these functions of state, in a fluid flow we also need the velocity **u** for a complete description. The velocity and the thermodynamic variables are functions of position **r** and time $t$.

The equations of hydrodynamics essentially express the conservation of momentum, mass and often also energy and/or other quantities such as chemical components of the fluid. They are partial differential equations containing the time derivatives of the velocity and the thermodynamic variables. First of all, the application of Newton's second law to a fluid element of volume $\delta V$ gives us

$$\rho \, \delta V \frac{d\mathbf{u}}{dt} = \delta \mathbf{F}, \tag{1.2}$$

where $\rho$ is the density of the fluid and $\delta \mathbf{F}$ is the force on the fluid element. Dividing by $\delta V$ and splitting the right-hand side up into different types of force we have

$$\rho \frac{d\mathbf{u}}{dt} = \mathbf{F}_{surface} + \mathbf{F}_{body}$$
$$= -\nabla P + \mathbf{F}_{visc} + \rho \mathbf{g} + \text{etc}, \tag{1.3}$$

where the terms on the right-hand side now represent various forces per unit volume. These forces fall into two classes. First there are *surface forces*, where the force on a fluid element comes from its immediate neighbours: the pressure gradient force,

---

[3] The letters and symbols used in these equations and others in this book are listed in appendix A.4.

present in all fluids, and the viscous force. One can consider that the pressure is defined (apart from some additive constant) by this equation. Alternatively, pressure is defined in a non-viscous fluid as the force per unit area exerted by a fluid element on its neighbours; the net force per unit volume appearing above is found by equating $\oint P \, d\mathbf{S} = \int \nabla P \, dV$. In a viscous fluid the force exerted by an element on its neighbours is generally not the same in all directions and the average is not necessarily equal to $P$; the definition of pressure in this case is less straightforward —see chapter 4.

Secondly, there are *body forces* such as gravity (**g** is the local gravitational force per unit mass). Depending on the context, there can be various other body forces. For instance, in chapter 7 we look at effects of the Coriolis force present in any fluid in a rotating frame of reference. In ionised gases there is also generally an electromagnetic body force (chapter 8).

Note that the derivative on the left-hand side of (1.3), $d/dt$, is the *Lagrangian* (co-moving) time derivative, which is related to the *Eulerian* (fixed in space) time derivative $\partial/\partial t$ in the following way. First recall that an infinitesimal change $\delta f$ in a function $f(x, y, z, t)$ can be expressed as

$$\delta f = \delta t \left(\frac{\partial f}{\partial t}\right)_{x,y,z} + \delta x \left(\frac{\partial f}{\partial x}\right)_{y,z,t} + \delta y \left(\frac{\partial f}{\partial y}\right)_{x,z,t} + \delta z \left(\frac{\partial f}{\partial z}\right)_{x,y,t}. \tag{1.4}$$

Dividing this equation by $\delta t$ and recognising that $u_x = \delta x/\delta t$ and so on, we can express the rate of change of any quantity $f(\mathbf{r}, t)$ in a fluid element moving with velocity **u** as

$$\frac{df}{dt} = \frac{\partial f}{\partial t} + \mathbf{u} \cdot \nabla f; \tag{1.5}$$

in other words, the co-moving rate of change of a quantity in a particular fluid element momentarily located at **r** is equal to the rate of change fixed at that location **r** plus the spatial derivative in the direction of the fluid velocity multiplied by the magnitude of the fluid velocity. Note that often a capital D is used for the Lagrangian derivative instead of d.

Now we use conservation of mass to derive the second equation. Imagining a volume $V$ with boundary $S$ (fixed in space), the rate of change of mass in the volume is equal to the mass flux $\rho\mathbf{u}$ into the volume through the boundaries, giving

$$\frac{\partial}{\partial t} \int \rho \, dV = - \oint \rho\mathbf{u} \cdot d\mathbf{S}, \tag{1.6}$$

$$= - \int \nabla \cdot (\rho\mathbf{u}) dV, \tag{1.7}$$

where the boundary element vector $d\mathbf{S}$ is defined pointing outwards; the second line follows from Gauss' theorem (A.15). Taking the time derivative inside the integrand,

and noting that this relation is valid for any volume $V$, gives us the continuity equation

$$\frac{\partial \rho}{\partial t} = -\nabla \cdot (\rho \mathbf{u}). \tag{1.8}$$

Often, we wish to have this equation in a form containing the Lagrangian derivative. Using (1.5) and (A.11), we have

$$\frac{d\rho}{dt} = -\rho \nabla \cdot \mathbf{u}. \tag{1.9}$$

So far, we have two equations (1.3) and (1.8) and three unknowns $\mathbf{u}$, $\rho$ and $P$. To close this set, the simplest option is to find some way of directly relating $\rho$ to $P$ without involving any new variables. This is known as a 'barotropic' equation of state where $\rho = \rho(P)$. A special case is to assume a constant density: $\rho = $ const, so that $\rho$ can be replaced by a constant $\rho_0$ in (1.3) and then (1.8) reduces to $\nabla \cdot \mathbf{u} = 0$ (see section 2.3.1 for more detail on incompressible fluids). However, we often have two or more independent thermodynamic variables and the equation of state of the fluid is expressible as $\rho = \rho(P, X_1, X_2...)$ where $X_1$, $X_2$ are some other thermodynamic variables.

We often have two thermodynamic degrees of freedom, or in other words, two independent functions of state, in which case we need one more partial differential equation in addition to the momentum equation and continuity equation derived above. An example of a fluid with two thermodynamic degrees of freedom would be an ideal gas with constant molar mass, as is used often in this book. In this and many other fluids the last partial differential equation describes the evolution of heat energy.

There are various ways of writing down the energy equation; what they all have in common is that they give the time derivative of a thermodynamic variable other than density. Now, we know from the first and second laws of thermodynamics that $dU = dQ + dW$ and if the change is reversible[4] that $dQ = T \, dS$ and $dW = -P \, dV$. This suggests a simple way of writing down the energy equation:

$$\frac{ds}{dt} = \frac{q}{T}, \tag{1.10}$$

where $s$ is specific entropy and $q$ is heat energy deposited per unit mass of fluid per unit time from an as yet unspecified source. This source could be external (for instance radiative heating or cooling), from neighbours via thermal diffusion, or internal (for instance from chemical or nuclear reactions, viscous heating, dissipation of electric currents).

---

[4] We are approximately in the reversible regime here because changes in volume are reversible, meaning that a given fluid element is pushing against neighbouring elements with the same pressure at which the neighbouring elements are pushing back, rather than the fluid element being allowed to expand into a neighbouring vacuum, for instance, where $-P \, dV$ is non-zero but work $dW = 0$.

Equation (1.10) on its own does not complete the set, indeed it introduces two new variables. In the *prognostic* equations (those with a time derivative on the left-hand side) we predict $\rho$ and $s$ so we need to find pressure and temperature with the *diagnostic* equations $P = P(\rho, s)$ and $T = T(\rho, s)$ in order to use them on the right-hand side of the prognostic equations. Depending on the fluid, these functions might be complicated or inconvenient so that it makes more sense to forgo (1.10) in favour of a form which does not involve entropy. This is normally the case with an ideal gas, for instance, when we have a simple equation of state relating $P$, $\rho$ and $T$ but more complex relations with $s$.

Searching for a more convenient alternative we can look at the equation for a reversible change $dU = dQ - P\,dV$ and write it down per unit mass of the fluid, which has a volume $1/\rho$ and internal energy $\epsilon$, and per unit time:

$$\frac{d\epsilon}{dt} = q - P\frac{d}{dt}(1/\rho)$$
$$= q - \frac{P}{\rho}\nabla \cdot \mathbf{u}, \qquad (1.11)$$

where (1.9) has been used. Since a new variable $\epsilon$ has been introduced and our independent variables now are $\rho$ and $\epsilon$, we also need the function $P = P(\rho, \epsilon)$ so we can plug pressure into the right-hand sides of (1.3) and (1.11).

In the case of an ideal gas this function is $P = (\gamma - 1)\rho\epsilon$ where $\gamma = c_p/c_v$ is the ratio of specific heats; with some rearrangement the energy equation can now be written

$$\frac{dP}{dt} = (\gamma - 1)\rho q - \gamma P \nabla \cdot \mathbf{u}. \qquad (1.12)$$

So in summary we have partial differential equations containing time derivatives of the velocity $\mathbf{u}$ and the independent thermodynamic functions of state. In the next chapter, rather than trying to solve these general equations we shall make the simplifications that we have no diffusion of momentum or heat, and no additional heating from any source, i.e. $q = 0$ and $\mathbf{F}_{\text{visc}} = \mathbf{0}$.

## Exercise

### 1.1 Different forms of the energy equation

Using the ideal gas relations $P = (\gamma - 1)\rho\epsilon$ and $P = \rho R_\mu T$, derive the energy equation (1.12) from (1.11) and furthermore show that the energy equation can also be written

$$(\gamma - 1)\frac{dT}{dt} = \frac{q}{R_\mu} - T \nabla \cdot \mathbf{u}. \qquad (1.13)$$

# Chapter 2

## Some basic concepts

We shall now look at some important concepts and tools to help us in the following chapters. To make this a little simpler, we neglect viscosity everywhere in this chapter—we have a so-called *ideal fluid*. Using the results from chapter 1, we can write down the equations governing the motion of an ideal fluid in a system containing gravity as an example of a body force:

$$\frac{\partial \mathbf{u}}{\partial t} + (\mathbf{u} \cdot \nabla)\mathbf{u} = -\frac{1}{\rho}\nabla P + \mathbf{g}; \tag{2.1}$$

$$\frac{\partial \rho}{\partial t} = -\nabla \cdot (\rho \mathbf{u}); \tag{2.2}$$

$$\frac{\mathrm{d}s}{\mathrm{d}t} = 0. \tag{2.3}$$

These are often called the Euler equations. The first equation is obtained by dividing equation (1.3) by density and breaking up the time derivative on the left with the help of (1.5). In the third equation, $s$ is the specific entropy; this form may be obtained by setting $q = 0$ in (1.10).

## 2.1 Visualisation

When working with three-dimensional vector fields, visualisation is half of the work. A few different types of line are commonly used to visualise a flow: streamlines, pathlines, streaklines and timelines:

- *Streamlines* are lines which are everywhere tangential to the velocity. They are conceptually similar therefore to electric or magnetic field lines—they illustrate the direction of the velocity field at a particular point in time. A *streamtube* is simply a tube bounded by streamlines (analogous to a flux tube in magnetohydrodynamics). Streamlines and streamtubes contrast with the

doi:10.1088/978-1-6817-4597-8ch2

following kinds of line, which give us information about what the flow was doing before the present time.

- A *pathline* (or *particle path*) is a line that goes through the locations a particular particle (or fluid element) has visited in the past. Normally therefore, one end of the line is the present location of the particle in question. In a transparent fluid one can imagine inserting some tiny lamps (or willing hovering fireflies) to be carried along passively with the flow. A long photographic exposure would show pathlines.

- A *streakline*, in contrast, joins fluid elements which have passed through a particular location at some time in the past. Experimentally, one can see a streakline by continuously injecting dye into the flow at a particular location.

- A *timeline* connects an arbitrary string of fluid elements at a particular point in time. One visualises with the aid of a family of timelines, each passing through the same fluid elements at different times. Alternatively one can imagine plotting the position of a co-moving stretchable ribbon at different times.

The various kinds of line are illustrated in figure 2.1.

At this juncture it is useful to consider the difference between *steady* and *unsteady* *flow*. As the name suggests, a steady flow has $\partial/\partial t = 0$ for all variables. (Important to note of course is that this does not mean that $d/dt = 0$.) It is useful to convince oneself that in a steady flow, streamlines, pathlines and streaklines are all the same. Chapter 3 is dedicated to steady flows.

## 2.2 Sound waves

There are many kinds of waves that can propagate through fluids; a small selection of these are covered in chapters 5, 7 and 8. The large variety comes from the fact that all waves and oscillations require a restoring force, and that many different restoring forces (and combinations of restoring forces) are available, such as gravity, Coriolis force, Lorentz force, viscoelasticity. There is only one kind of wave, however, which

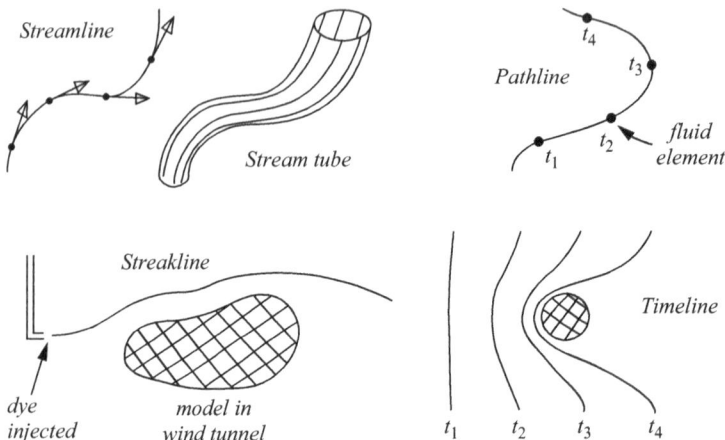

**Figure 2.1.** Streamlines, streamtubes, pathlines, streaklines, and timelines.

can exist in *all* fluids: the sound wave. The reason for this is that its restoring force is the pressure gradient force—the term with $-\nabla P$ in the momentum equation (2.1)—which is the only force present in all fluids.

An important first step when examining the properties of a wave is to derive the dispersion relation, the relation between frequency and wavelength, or rather between angular frequency $\omega$ and wavenumber $k$. We shall now do this for sound waves. Consider a homogenous system without gravity where the background state is constant in space and there is no background fluid flow. The sound wave causes the variables (density, pressure, velocity) to oscillate around this equilibrium. Furthermore the variables vary only in the $x$ direction, so that any wave must also propagate in that direction (we call this a plane wave), and the velocity in the $x$ direction is $u$. The momentum and mass conservation equations (2.1) and (2.2) can be written:

$$\frac{\partial u}{\partial t} + u\frac{\partial u}{\partial x} = -\frac{1}{\rho}\frac{\partial P}{\partial x}, \tag{2.4}$$

$$\frac{\partial \rho}{\partial t} = -\frac{\partial}{\partial x}(\rho u). \tag{2.5}$$

In order to look at the properties of low-amplitude, linear waves, it is necessary to linearise these equations. To do this, we first define some background equilibrium state with pressure $P_0$, density $\rho_0$ and zero velocity, and write pressure $P = P_0 + \delta P$, $\rho = \rho_0 + \delta\rho$ where $\delta P \ll P_0$ and $\delta\rho \ll \rho_0$. It follows that the velocity $u$ is also small. Now, looking at the ratio of the two terms on the left-hand side of (2.4), we see that we can ignore the second provided that

$$\frac{U}{T} \gg \frac{U^2}{L} \quad \Rightarrow \quad L \gg A, \tag{2.6}$$

where $T$ is the timescale (the wave period), $L$ is the length scale (the wavelength) and $A \sim UT$ is the amplitude (how far a fluid element moves from its equilibrium position). This condition is easily shown to be equivalent to the conditions $\delta P \ll P_0$ and $\delta\rho \ll \rho_0$ introduced above.

In addition we need some relation between the pressure and density perturbations: we define $c^2 \equiv \partial P/\partial\rho$. The linearised equations are

$$\frac{\partial u}{\partial t} = -\frac{1}{\rho_0}\frac{\partial(\delta P)}{\partial x}, \tag{2.7}$$

$$\frac{\partial(\delta\rho)}{\partial t} = -\rho_0\frac{\partial u}{\partial x}, \tag{2.8}$$

$$\delta P = c^2\delta\rho. \tag{2.9}$$

Substituting for $\delta\rho$ from (2.9) into (2.8), differentiating with respect to $t$ and combining with (2.7) gives the wave equation

$$\frac{\partial^2(\delta P)}{\partial t^2} = c^2\frac{\partial^2(\delta P)}{\partial x^2}.\tag{2.10}$$

The general solution to this is

$$\delta P = A(x - ct) + B(x + ct).\tag{2.11}$$

The two terms represent waves travelling in the positive and negative $x$-directions, respectively. Considering a wave travelling either to the right or to the left, whatever form of $\delta P(x)$ is present at time $t = 0$ it is preserved in shape but is shifted a distance $ct$ at a later time $t$, i.e. it moves with speed $c$.

Alternatively we could have dealt with (2.10) by assuming a solution of the form $\delta P = Ae^{i(kx-\omega t)}$, giving a dispersion relation of

$$\omega = kc.\tag{2.12}$$

Recall the definitions of phase speed $\omega/k$, the speed at which individual peaks and troughs move, and group speed $\partial\omega/\partial k$, the speed at which wavepackets and information move. In this case, both are equal to $c$, which does not depend on the frequency. This non-dependence of speed on frequency we call *non-dispersiveness*. We return to this topic in chapter 5.

Finally, how do we calculate the sound speed? It was introduced as the square root of the ratio between the density and pressure perturbations. Clearly this is only valid for small perturbations. Its value is normally calculated by assuming that the motion is adiabatic and the fluid elements have constant entropy. In an ideal gas, this gives an (adiabatic) sound speed

$$c = \sqrt{\frac{\gamma P}{\rho}} = \sqrt{\frac{\gamma RT}{\mu_{\mathrm{m}}}},\tag{2.13}$$

where the second form is useful because it shows the dependence on temperature if $\gamma$ and $\mu_{\mathrm{m}}$ are constant. Table 2.1 lists approximate sound speeds in various materials.

**Table 2.1.** The sound speed in various relevant media.

| Medium | Sound speed (km s$^{-1}$) |
| --- | --- |
| Rubber | 0.1 |
| Air at room temperature | 0.3 |
| Water | 1.5 |
| Interior of Earth | 8 |
| Interstellar medium at $10^4$ K | 10 |
| Interior of the Sun | 300 |
| Interior of a (cold) white dwarf | 5000 |
| Interior of a (cold) neutron star | $10^5$ |
| Relativistic non-degenerate plasma | $c_{\mathrm{light}}/\sqrt{3}$ |

Note finally that occasionally the *isothermal* sound speed is discussed, which in an ideal gas would be $c_{ith} = \sqrt{P/\rho}$. This can be relevant in contexts with long wave periods and efficient heat transfer.

## 2.3 Compressibility

We can estimate the fractional density variation in a flow in the following way. Imagine the system has characteristic length and time scales $L$ and $T$ and typical velocity $U$. The sound speed is $c^2 = (\partial P/\partial \rho)_s$ (see section 2.2) and the Mach number is defined as $M \equiv U/c$. In a system without gravity and other body forces, comparing the sizes of the terms in the momentum equation (2.1) gives:

$$\frac{\partial \mathbf{u}}{\partial t} + (\mathbf{u} \cdot \nabla)\mathbf{u} = -\frac{1}{\rho}\nabla P$$

$$\frac{U}{T} \qquad \frac{U^2}{L} \qquad \frac{\delta P}{\rho L} \tag{2.14}$$

$$M^2\frac{L}{UT} \qquad M^2 \qquad \frac{\delta \rho}{\rho},$$

where $\delta P$ is the typical departure of the pressure from some mean or equilibrium pressure. The third line is obtained by multiplying the second by $L/c^2$. In a general unsteady flow we have $L/T \sim U$ so the first two terms will be of comparable size and we have

$$\frac{\delta \rho}{\rho} \sim M^2. \tag{2.15}$$

This relation also holds for steady flows (as discussed in chapter 3), where all Eulerian time derivatives $\partial/\partial t$ are zero and the first term on the left of (2.14) is absent. However, in the special case of a stationary fluid hosting a sound wave (as we saw in section 2.2) we have $L/T \sim c$ instead of $L/T \sim U$, the second term on the left can be neglected and instead of (2.15) we have

$$\frac{\delta \rho}{\rho} \sim M. \tag{2.16}$$

In any case, this demonstrates how density changes within the fluid arise through the inertia of the fluid, and that if the motion is subsonic, i.e. if $M$ is small, then the fractional density differences $\delta\rho/\delta$ are small. In the literature it is customary therefore to say that the flow is incompressible if $M < 0.3$ so that relative density variations are less than a tenth.

### 2.3.1 The incompressible equations of motion

If a flow is approximately incompressible we can replace the density $\rho$ in the momentum equation (2.1) with a constant density $\rho_0$:

$$\frac{\partial \mathbf{u}}{\partial t} + (\mathbf{u} \cdot \nabla)\mathbf{u} = -\frac{1}{\rho_0}\nabla P \qquad (2.17)$$

and the continuity equation (2.2) with

$$\nabla \cdot \mathbf{u} = 0. \qquad (2.18)$$

In principle this completes the incompressible set of equations because we have the same number of equations as variables. It is not obvious, however, how to find $P$. The trick is to take the divergence of the momentum equation which, using (2.18), gives the Laplace equation

$$\nabla^2 P = 0, \qquad (2.19)$$

which is a boundary value problem. Formally, what we have done here is to set the sound speed $(\partial P / \partial \rho)_s$ to infinity, so that any pressure perturbation is immediately transmitted to the entire volume.

That these equations do not allow sound waves can be very advantageous. It is generally helpful to have equations which only allow the phenomenon we are interested in, as it makes it easier to find the solution we are interested in. In numerical work, the timestep is set by the shortest timescale present in the system, which is often the acoustic timescale. Filtering out the sound waves allows a longer timestep than would be possible with the full compressible equations.

We shall use the incompressible equations in chapter 5 in deriving dispersion relations for water waves. They are also used in many other contexts such as in (subsonic) aerodynamics and the flow of liquids, e.g. of water through a pipe.

In general and in various contexts, approximations are useful and there are almost countless such approximations in the literature. In section 2.5 we shall look at some further approximations that are useful if we have a constant gravity force $\mathbf{g}$ in the momentum equation.

## 2.4 Rotation of fluid elements

To clarify, in this section we shall examine the rotation of fluid elements about their own centres of mass, as opposed to motion in a circle around some external axis. (Both are covered in more detail in chapter 7.)

### 2.4.1 Circulation

Let us define the circulation $\Gamma$ around a closed co-moving loop as

$$\Gamma \equiv \oint \mathbf{u} \cdot \delta\mathbf{s}, \qquad (2.20)$$

where $\delta\mathbf{s}$ is the infinitesimal displacement along the loop. As the loop moves with the flow, the rate of change of this circulation is

$$\frac{d\Gamma}{dt} = \frac{d}{dt} \oint \mathbf{u} \cdot \delta\mathbf{s}$$

$$= \oint \frac{d\mathbf{u}}{dt} \cdot \delta\mathbf{s} + \oint \mathbf{u} \cdot \frac{d}{dt}(\delta\mathbf{s})$$

$$= \int \nabla \times \frac{d\mathbf{u}}{dt} \cdot \delta\mathbf{S} + \oint \mathbf{u} \cdot \delta\left(\frac{d\mathbf{s}}{dt}\right)$$

$$= \int \nabla \times \left(-\frac{1}{\rho}\nabla P + \mathbf{g}\right) \cdot \delta\mathbf{S} + \oint \mathbf{u} \cdot \delta\mathbf{u} \tag{2.21}$$

$$= \int \left(\frac{1}{\rho^2}\nabla P \times \nabla\rho\right) \cdot \delta\mathbf{S} + \oint \frac{1}{2}\delta(u^2),$$

$$= \int \left(\frac{1}{\rho^2}\nabla P \times \nabla\rho\right) \cdot \delta\mathbf{S},$$

where Stokes' theorem and some vector calculus identities have been used (see appendix A.3), and where $\delta\mathbf{S}$ is an element of some surface bounded by the loop. In getting from the third to the fourth line, the momentum equation (2.1) is used, including gravity as the sole body force, which prompty disappears because its curl is zero (it is a *conservative* force). In the last step, the second term on the right is dropped as it is equal to zero.

We now digress to introduce two words used often in hydrodynamics. A flow is *baroclinic* if the surfaces of constant pressure intersect the surfaces of constant density, or in other words, if $\nabla P$ and $\nabla\rho$ are not parallel. In such a flow, we have two degrees of freedom in the thermodynamic state of the fluid since pressure and density can be varied independently of each other. In any case, the remaining term in (2.21) could be called a baroclinic term, as it is non-zero only in a baroclinic flow.

If a flow is not baroclinic then it is *barotropic* (see figure 2.2). In a barotropic flow $\nabla P$ and $\nabla\rho$ are parallel and we can write $\rho = \rho(P)$; there is only one thermodynamic degree of freedom. It is important to note (since there is a lot of confusion about this in the literature) that a barotropic flow is not the same thing as the flow of a fluid with a barotropic equation of state. A fluid with a barotropic equation of state $\rho = \rho(P)$ necessarily flows barotropically, but a fluid with a more complex equation

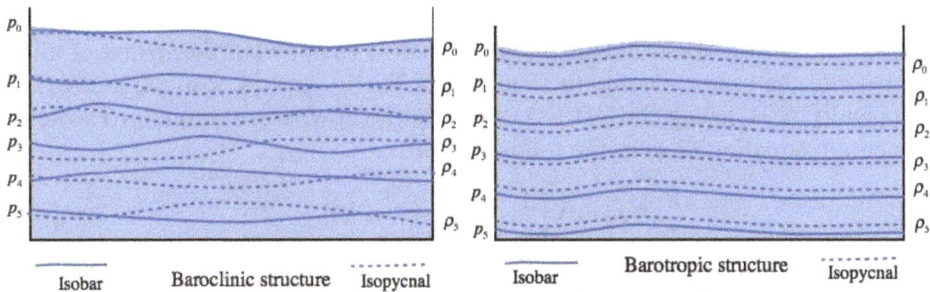

**Figure 2.2.** The difference between baroclinic and barotropic flow.

of state, e.g. $\rho = \rho(P, T)$, can also flow barotropically under the right conditions, for instance if heat transfer is so efficient that temperature can be considered constant, or if heat transfer is negligible and specific entropy is constant.

Anyway, we see that in an inviscid barotropic flow, circulation is a conserved quantity:

$$\frac{d\Gamma}{dt} = 0. \tag{2.22}$$

This is known as *Kelvin's circulation theorem*. Circulation can be generated and destroyed via viscosity, baroclinicity or non-conservative body forces (i.e. with body-forces with non-zero curl).

### 2.4.2 Vorticity

The *vorticity* is simply the curl of velocity:

$$\omega \equiv \nabla \times \mathbf{u}. \tag{2.23}$$

Imagine a small spherical fluid element of radius $a$ which is rotating with angular velocity $\mathbf{\Omega}$. Considering now the circulation of a loop around the 'equator' of this element and using Stokes' theorem to relate this to vorticity, we see

$$\Gamma = \oint \mathbf{u} \cdot d\mathbf{s} = \int \omega \cdot d\mathbf{S} \tag{2.24}$$

$$2\pi a^2 \Omega = \pi a^2 \omega$$
$$2\Omega = \omega, \tag{2.25}$$

where the surface integral is taken over the equatorial plane, which is perpendicular to the vorticity. So the vorticity is simply double the angular velocity of a fluid element.

### 2.4.3 Potential flow

Looking at (2.22) and (2.24) leads us to the following statement concerning inviscid barotropic flow: if at some point in time the vorticity vanishes at every location, then it also vanishes at all other times. Such a flow is consequently called *irrotational*.

This is useful because we can simplify the equations of an irrotational flow by expressing the curl-free velocity field as the gradient of a scalar: $\mathbf{u} = \nabla\phi$. Consequently irrotational flow is often referred to as *potential flow*. First of all we make use of the vector identity $(\mathbf{u} \cdot \nabla)\mathbf{u} = (1/2)\nabla(u^2) - \mathbf{u} \times (\nabla \times \mathbf{u})$ (equation (A.8) in appendix A.3), losing the last term because the flow is irrotational, to write the momentum equation (2.1) in the following form:

$$\frac{\partial}{\partial t}(\nabla\phi) + \nabla\left(\frac{1}{2}u^2\right) = -\nabla h - \nabla\Phi, \tag{2.26}$$

where $\Phi$ is the gravitational potential (so that $\mathbf{g} = -\nabla\Phi$) and the pressure gradient term has been reorganised with the help of a new function of state $h = h(P)$ defined by $\nabla h \equiv (1/\rho)\nabla P$, made possible by the barotropic condition $\rho = \rho(P)$, i.e. that there is only one independent thermodynamic variable.

Collecting terms we have

$$\frac{\partial \phi}{\partial t} + \frac{1}{2}u^2 + h + \Phi = f(t). \tag{2.27}$$

In an unsteady irrotational inviscid barotropic flow the sum of the four terms on the left is constant in space but not in time, and so can be set equal to some function $f(t)$ of time but not of location.

This equation will be of great use to us when we come to analyse waves in chapter 5.

## 2.5 Gravitationally stratified fluid

In many contexts a fluid is pulled in one direction by an external gravitational force, which appears in the momentum equation as **g** as in (2.1). To prevent the fluid accelerating uncontrollably, gravity is balanced by a pressure gradient. This is called hydrostatic equilibrium.

### 2.5.1 Hydrostatic equilibrium

In many situations the fluid is completely stationary, and we need therefore

$$0 = -\nabla P + \rho\mathbf{g}, \tag{2.28}$$

which comes simply from setting velocity to zero in (2.1). Taking gravity to be directed downwards and the $z$ axis to be pointing upwards, we have the well-known equation of *hydrostatic equilibrium*

$$\frac{\partial P}{\partial z} = -\rho g. \tag{2.29}$$

This relation can be used, for instance, to determine the structure of the atmosphere. In an isothermal gas atmosphere this equation is easily integrated, using the equation of state $P = \rho R_\mu T$ (where $R_\mu \equiv R/\mu_{\mathrm{m}}$, which is often used where mean molecular weight $\mu_{\mathrm{m}}$ is constant). The solution is

$$P = P_0 \exp\left(-\frac{z}{H_p}\right) \quad \text{where} \quad H_p \equiv \left(\frac{\partial \ln P}{\partial z}\right)^{-1} = \frac{R_\mu T}{g}, \tag{2.30}$$

where the above definition of the *pressure scale height* $H_p$ is valid also for non-isothermal atmospheres. In the Earth's atmosphere, its value is around 8 km, roughly the height of Mount Everest. A neutron-star atmosphere may have a scale height as little as 1 cm, which is probably the smallest macroscopic scale found anywhere in astrophysics. In contrast the gas in a Galaxy cluster, which is also thought to be in approximate hydrostatic equilibrium, has a scale height of order megaparsecs (see appendix A.2 for a list of astrophysical units). Finally note that in addition to the pressure scale height, the *density scale height* $H_\rho \equiv (\partial \ln \rho/\partial z)^{-1}$ is sometimes discussed; in gases the two are usually comparable to one another but in liquids such as the ocean the two scale heights have quite different values.

The Earth's atmosphere is *approximately* in hydrostatic equilibrium—the vertical motion we observe is the result of a relatively tiny inbalance between gravity and the vertical pressure gradient force. Such conditions also prevail in the oceans, stars, accretion discs, the interior of the Earth, teacups and bowls of miso soup. Below, we look at various simplifications that can be made to the equations in these situations.

### 2.5.2 The Boussinesq and anelastic approximations

Often in gravitationally stratified fluids we are interested in modelling phenomena which take place on a timescale much longer than the acoustic timescale and can make simplifications to the full equations (2.1)–(2.3) to filter out sound waves. In this section we look at the Boussinesq and anelastic approximations. The former is named after the Frenchman Joseph Boussinesq who first applied it at the end of the 19th century; the latter's name refers to the fact that the fluid is not allowed to have elastic energy, such as the compression/rarefaction energy present in sound waves.

These approximations are both used for gravitationally stratified fluids where the motion is subsonic, and where the pressure and density at any given height vary only slightly from reference equilibrium values at that height, so we can write $P = P_0(z) + \delta P$ and $\rho = \rho_0(z) + \delta\rho$. Now we can play with the vertical component of the momentum equation (2.1), dropping second-order terms in $\delta\rho/\rho$:

$$\frac{dw}{dt} = -\frac{1}{\rho_0 + \delta\rho}\frac{\partial(P_0 + \delta P)}{\partial z} - g$$

$$\frac{dw}{dt} \approx -\frac{1}{\rho_0}\left(\frac{\partial P_0}{\partial z} + \frac{\partial \delta P}{\partial z} - \frac{\delta\rho}{\rho_0}\frac{\partial P_0}{\partial z}\right) - g,$$

at which point we can substitute from the equilibrium condition $\partial P_0/\partial z = -\rho_0 g$ to obtain

$$\frac{dw}{dt} = -\frac{1}{\rho_0}\left(\frac{\partial \delta P}{\partial z} - g\,\delta\rho\right). \tag{2.31}$$

The horizontal components of the momentum equation can also be simplified somewhat from their original form in (2.1) by replacing $\rho$ with $\rho_0$. The momentum equation including all three directions then becomes

$$\frac{d\mathbf{u}}{dt} = -\frac{1}{\rho_0}\nabla\,\delta P + \frac{\delta\rho}{\rho_0}\mathbf{g}. \tag{2.32}$$

So far, the Boussinesq and anelastic approximations are the same; the difference will become clear when we look at the continuity equation $\partial\rho/\partial t + \mathbf{u}\cdot\nabla\rho = -\rho\nabla\cdot\mathbf{u}$ and the sizes of its various terms:

$$\frac{\partial\,\delta\rho}{\partial t} \;+\; \mathbf{u}\cdot\nabla\delta\rho \;+\; \mathbf{u}\cdot\nabla\rho_0 \;=\; -\rho_0\nabla\cdot\mathbf{u} \;-\; \delta\rho\nabla\cdot\mathbf{u} \tag{2.33}$$

$$
\begin{array}{ccccc}
\dfrac{\delta\rho}{T} & \left(\dfrac{U}{L_h}\text{ or }\dfrac{W}{L_z}\right)\delta\rho & \dfrac{W\rho_0}{H_\rho} & \left(\dfrac{U}{L_h}\text{ or }\dfrac{W}{L_z}\right)\rho_0 & \left(\dfrac{U}{L_h}\text{ or }\dfrac{W}{L_z}\right)\delta\rho \\[1.2em]
\dfrac{\delta\rho}{\rho_0} & \dfrac{\delta\rho}{\rho_0} & \dfrac{L_z}{H_\rho} & 1 & \dfrac{\delta\rho}{\rho_0}
\end{array}
\tag{2.34}
$$

where in getting from the second to the third line it is assumed that the flow timescale $T$, horizontal and vertical length scales $L_h$ and $L_z$, and velocities $U$ and $W$ are related by $1/T = U/L_h = W/L_z$, and everything is then multiplied by $T/\rho_0$. We have already assumed that $\delta\rho \ll \rho_0$ so the first, second and fifth terms can be dropped.

If $L_z \ll H_\rho$, the third term can also be dropped and only the fourth term remains. If however this is not the case, we must keep the third term. This is the crux of the difference between the Boussinesq and anelastic approximations. The continuity equation in these two cases can be simplified to the following:

$$\text{Boussinesq:} \quad \nabla\cdot\mathbf{u} = 0 \tag{2.35}$$

$$\text{anelastic:} \quad \nabla\cdot\rho_0\mathbf{u} = 0. \tag{2.36}$$

This still leaves us needing one more equation to complete the set, because unlike in the incompressible case as described in section 2.1, the momentum equation (2.32) contains $\delta\rho$ as an independent variable. As with the equations derived in section 1.2, there is more than one possible way to do this. First though it is useful to examine the relative sizes of $\delta P/P_0$ and $\delta\rho/\rho_0$; assuming approximate hydrostatic equilibrium,

$$\frac{\partial P_0}{\partial z} = -\rho_0 g \;\text{ and }\; \frac{\partial\,\delta P}{\partial z} \approx -\delta\rho g \;\Rightarrow\; \frac{\delta\rho}{\rho_0} \approx \frac{\dfrac{\partial\,\delta P}{\partial z}}{\dfrac{\partial P_0}{\partial z}} \sim \frac{\delta P}{P_0}\cdot\frac{H_\rho}{L_z}. \tag{2.37}$$

We see then that in the Boussinesq case where $L_z \ll H_\rho$, we have $\delta P/P \ll \delta\rho/\rho$[1].

Given that we have two thermodynamic degrees of freedom, we can replace the variation in density $\delta\rho$ in the momentum equation (2.32) with the variations in two other functions of state. One of those should be pressure $\delta P$ because it is already in the equations. Obvious choices for the other are temperature and entropy, for which we can write down energy equations containing the relevant thermal physics and thereby complete the set of equations. Recalling that the fluctuations about the reference state are small, we can write (with the help of some relations the reader can find in appendix A.5)

---

[1] Albeit with the caveat that we have been a bit sloppy with the difference between $H_p$ and $H_\rho$. Fortunately this seems to be a victimless crime.

$$\frac{\delta\rho}{\rho_0} = \frac{1}{\rho_0}\left[\left(\frac{\partial\rho}{\partial T}\right)_P \delta T + \left(\frac{\partial\rho}{\partial P}\right)_T \delta P\right]$$
$$= -\alpha\,\delta T + \kappa\,\delta P \tag{2.38}$$

if we want an energy equation with temperature, and if we prefer entropy we can write

$$\frac{\delta\rho}{\rho_0} = \frac{1}{\rho_0}\left[\left(\frac{\partial\rho}{\partial s}\right)_P \delta s + \left(\frac{\partial\rho}{\partial P}\right)_s \delta P\right]$$
$$= -\frac{\alpha T_0}{c_p}\delta s + \frac{\kappa}{\gamma}\delta P, \tag{2.39}$$

where $\alpha \equiv -(1/\rho)(\partial\rho/\partial T)_P$ is the coefficient of thermal expansion and $\kappa \equiv (1/\rho)(\partial\rho/\partial P)_T$ is the coefficient of isothermal compressibility (see appendix A.5). For an ideal gas $\alpha T = 1$ and $\kappa P = 1$, and for water at room temperature and pressure $\alpha T \approx 0.1$ and $\kappa P \approx 5 \times 10^{-5}$, but note that water below 4 °C has negative $\alpha$.

In the Boussinesq system (but not in the anelastic) we then normally drop the terms with $\delta P$ in equations (2.38) and (2.39). This means that density variations are the result of temperature (or entropy) variations and motions are driven mainly by bouyancy; pressure is close to constant on any given horizontal plane. A good example is the flow of air around a room under the influence of heating from a radiator and cooling at a window. In other contexts, the role of temperature or entropy is replaced or joined by a different variable, such as salinity in the ocean. Anyway, an extra equation is still required to complete the set; in the case of air flow in a room we then need an energy equation which gives us the time derivative of temperature or entropy. For the energy equation with temperature, see exercise 1.2. Here we write down the equation for entropy:

$$\frac{ds}{dt} = \frac{\partial s}{\partial t} + \mathbf{u}\cdot\nabla s = \frac{\partial\,\delta s}{\partial t} + w\frac{\partial}{\partial z}(s_0 + \delta s) = \frac{q}{T}, \tag{2.40}$$

where $q$ is the specific heating rate and $w$ is the vertical component of velocity.

The Boussinesq approximation has been used in a myriad of contexts, and we use it in section 5.4 to investigate internal gravity waves. The anelastic approximation was invented by Ogura and Phillips (1962) and has been used extensively in atmospheric physics as well as to model the Earth's interior (e.g. Glatzmaier and Roberts 1996), stellar interiors (Browning, 2008), and solar convection (Kurowski et al 2014, Verhoeven et al 2015). A further discussion of the Boussinesq and anelastic approximations and their applicability can be found in Randall (2013) and Drake (2014).

### 2.5.3 The hydrostatic approximation

The two approximations in the previous sections require that the density field differs only slightly from some time-constant mean $\rho_0$. In many contexts they are therefore not applicable. Fortunately there is another method that works by assuming that (2.29) holds precisely; this equation then replaces the vertical component of the equation of motion. The logic here is that hydrostatic balance is established on the vertical acoustic timescale, which is much shorter than any timescales we wish to investigate, so that for practical purposes we are always in hydrostatic balance. In a fluid with two thermodynamic degrees of freedom (such as ideal gas with constant molar mass) the resulting set of equations then only contains three properly independent variables: the two horizontal components of velocity and one of the thermodynamic variables. The other thermodynamic variable is obtained by integrating (2.29) and the vertical component of velocity can be similarly obtained from the independent variables. The equations are not derived here as they are a bit lengthy (but no more difficult to solve than those in the approximations above); the reader is referred to Kasahara (1974).

This approximation filters out sound waves propagating in the vertical direction, but not in the horizontal direction. In numerical work it is therefore of no use in, say, the context of convection in a cloud or a bowl of miso soup because the timestep cannot be made any longer than with the full compressible equations; indeed in the horizontal directions there is little difference from the fully compressible equations. It is useful in contexts where the horizontal length scale is much greater than the vertical length scale. For instance, in the Earth's atmosphere the vertical acoustic timescale ($H_p/c$) is only 25 seconds—much smaller than any timescale we are interested in—whereas the horizontal acoustic timescale is about an hour if our characteristic length scale is 1000 km, which might be the case if we want to do

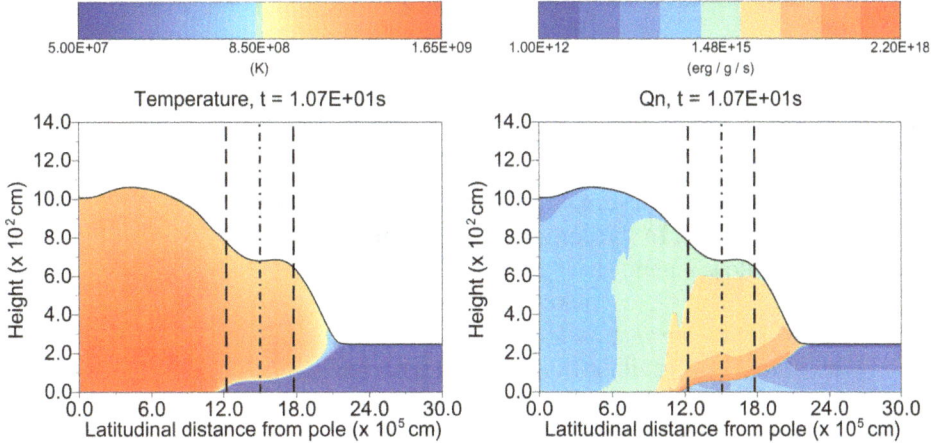

**Figure 2.3.** A snapshot from a simulation of a thermonuclear burst on the surface of a neutron star—helium accreted from a companion star burns into carbon. In this simulation the helium is ignited at the north pole (on the left-hand side) and the flame front propagates across the surface of the star, crossing the equator, towards the south pole. Plotted on the left is the temperature, which increases from $2 \times 10^8$ K to over $10^9$ K as the material burns. The scale height $H_p$ inceases correspondingly. On the right is plotted the specific nuclear heating rate $q_n$; the burning is concentrated in a slanted plane. Note the difference in horizontal and vertical length scales and the large deviation from the original equilibrium that make the hydrostatic approximation suitable but the Boussinesq and anelastic approximations unsuitable. (Reproduced with permission from Cavecchi *et al* 2013).

**Table 2.2.** Approximations and their conditions of applicability.

| Approximation | $L_z \ll L_h$ | $L_z \ll H_\rho$ | $u_h \ll c$ | $u_z \ll c$ | $\rho \approx \rho_0(z)$ | Poisson solver |
|---|---|---|---|---|---|---|
| Incompressible | no | — | yes | yes | yes | yes |
| Boussinesq | no | yes | yes | yes | yes | yes |
| Anelastic | no | no | yes | yes | yes | yes |
| Hydrostatic | yes | no | no | yes | no | no |

weather forecasting rather than model individual clouds. Using the hydrostatic approximation we can jump over the vertical acoustic timescale.

There are a variety of vertical coordinate systems available in this approximation: we can use a simple height coordinate $z$ but it is normally more convenient to use another coordinate such as pressure or entropy or some hybrid coordinate. For a full discussion the reader is referred to the review by Kasahara (1974). The hydrostatic approximation has also been extended to magnetohydrodynamics (Braithwaite and Cavecchi 2012), where it has been used to model thermonuclear bursts in neutron-star oceans (see figure 2.3).

In table 2.2, the four approximations covered in this chapter are summarised in terms of their requirements: whether the vertical length scale must be much less than either the horizontal length scale or the scale height, whether the horizontal or vertical velocities must be much less than the sound speed, or whether it is required that there are only small variations of density about some time-constant reference density $\rho_0(z)$, and a numerical requirement, namely whether a Poisson solver is needed (which is computationally awkward).

## Exercises

2.1. **Compressibility and energy**

In section 2.3 we estimated relative density variations in terms of the flow Mach number. In the context of a gas and with reference to thermal kinetic energy, flow kinetic energy and the work required for compression, explain how the result (2.15) makes intuitive sense. In light of this, the second result (2.16) appears *prima facie* impossible: somehow a greater compression can be produced (recalling that $M \ll 1$) with the same energy. Where does the energy required for the compression come from?

2.2. **Compressibility and gravity**

How does the conclusion of section 2.3, i.e. relation (2.15), change if we add a gravity force **g** as in (2.1)?

2.3. **Boussinesq symmetry**

Consider the air circulating in a room with an open window. The air inside can be either warmer or colder than outside. Explain why the flow in a cold room looks exactly the same as the flow in a warm room viewed upside down.

# References

Braithwaite J and Cavecchi Y 2012 A numerical magnetohydrodynamic scheme using the hydrostatic approximation *Mon. Not. R. Astron. Soc.* **427** 3265–79

Browning M K 2008 Simulations of dynamo action in fully convective stars *ApJ* **676** 1262–80

Cavecchi Y, Watts A L, Braithwaite J and Levin Y 2013 Flame propagation on the surfaces of rapidly rotating neutron stars during Type I x-ray bursts *Mon. Not. R. Astron. Soc.* **434** 3526–41

Drake J 2014 *Climate Modeling for Scientists and Engineers* (Philadelphia, PA: Society for Industrial and Applied Mathematics)

Glatzmaier G A and Roberts P H 1996 An anelastic evolutionary geodynamo simulation driven by compositional and thermal convection *Physica* D **97** 81–94

Kasahara A 1974 Various vertical coordinate systems used for numerical weather prediction *Mon. Weather Rev.* **102** 509–22

Kurowski M J, Grabowski W W and Smolarkiewicz P K 2014 Anelastic and compressible simulation of moist deep convection *J. Atmos. Sci.* **71** 3767–87

Ogura Y and Phillips N A 1962 Scale analysis of deep and shallow convection in the atmosphere *J. Atmos. Sci.* **19** 173–9

Randall D 2013 The anelastic and Boussinesq approximations http://kiwi.atmos.colostate.edu/group/dave/pdf/AneBous.pdf

Verhoeven J, Wiesehofer T and Stellmach S 2015 Anelastic versus fully compressible turbulent Rayleigh Bénard convection *ApJ* **805** 62

# Chapter 3

## Steady flow of an ideal fluid

In this chapter we look at various phenomena in steady flows, i.e. flows in which $\partial/\partial t = 0$. As in the previous chapter, we also make the simplification of ignoring viscosity, i.e. we consider an ideal fluid.

## 3.1 Bernoulli's equation

Let us consider a steady flow, i.e. a flow where $\partial/\partial t = 0$ (but the co-moving derivatives are in general non-zero). As mentioned previously, streamlines are lines everywhere tangential to the flow velocity, which in a steady flow are the same as pathlines. Consider a *streamtube*, i.e. a volume bounded by streamlines and at each end by a surface perpendicular to the flow—the fluid is flowing into the volume at one end and exiting at the other end (figure 3.1).

The rate of change of energy in this volume, which must vanish in a steady flow, is given by the difference between the energy entering and exiting and the difference between the $P \, dV$ work done at the ends of the tube on the fluid ahead of it. Denoting quantities at the inflow and outflow ends of the volume with subscripts 0 and 1, we can write this as

$$A_0 u_0 \varepsilon_0 \rho_0 - A_1 u_1 \varepsilon_1 \rho_1 + A_0 u_0 P_0 - A_1 u_1 P_1 = 0, \tag{3.1}$$

where $A$ and $u$ are the cross-sectional area of, and the velocity at, the ends, and $\varepsilon$ is total energy per unit mass—the sum of internal, kinetic and gravitational potential:

$$\varepsilon \equiv \epsilon + \frac{1}{2}u^2 + \Phi. \tag{3.2}$$

The mass in the volume must also be constant in time, so that $A_0 u_0 \rho_0 = A_1 u_1 \rho_1$. Substituted back into (3.1) this gives

$$\varepsilon_0 + \frac{P_0}{\rho_0} = \varepsilon_1 + \frac{P_1}{\rho_1}, \tag{3.3}$$

doi:10.1088/978-1-6817-4597-8ch3　　　　3-1

**Figure 3.1.** A streamtube in a steady flow. Fluid enters at pressure $P_0$ and density $\rho_0$ at speed $u_0$ and exits at pressure $P_1$ and density $\rho_1$ at speed $u_1$.

which is valid for any volume in the flow, so that we can say more generally that

$$\frac{d}{dt}\left(\varepsilon + \frac{P}{\rho}\right) = 0. \tag{3.4}$$

Physically this represents just conservation of energy. This condition holds only in an ideal fluid, where no energy can be transferred between neighbouring fluid elements *across* the streamlines since the component of the pressure gradient perpendicular to the streamlines is of course perpendicular to the velocity. (Note that unlike energy and mass, momentum *can* be transferred across streamlines.) In a viscous fluid this is no longer true: energy can be transferred by viscous stress; also heat conduction can transfer energy perpendicular to the flow.

This result can be derived in another way, providing us with a different, intuitive understanding of how it comes from energy conservation. The momentum equation (2.1) equates the acceleration of the fluid to the force per unit mass, and taking the dot product with **u** equates the rate of change of kinetic energy to the rate at which work is done by the various forces:

$$\mathbf{u} \cdot \frac{d\mathbf{u}}{dt} = -\mathbf{u} \cdot \left(\frac{1}{\rho}\nabla P\right) - \mathbf{u} \cdot \nabla\Phi,$$

$$\frac{d}{dt}\left(\frac{1}{2}u^2\right) = -\mathbf{u} \cdot \nabla\left(\frac{P}{\rho}\right) + P\mathbf{u} \cdot \nabla\left(\frac{1}{\rho}\right) - \frac{d\Phi}{dt}, \tag{3.5}$$

where the pressure gradient term has been broken up into two parts, the first of which is simply (minus) the Lagrangian time derivative of $P/\rho$; the second part is pressure times the Lagrangian derivative of specific volume, which in an ideal fluid (where there is no viscous or other heating and no conduction of heat) can be equated to the rate of $P\,dV$ work done on a unit mass—see (1.11). The gravity term is simply the Lagrangian derivative of the specific potential energy. Collecting terms we have Bernoulli's equation:

$$\frac{d}{dt}\left(\frac{1}{2}u^2 + \frac{P}{\rho} + \varepsilon + \Phi\right) = 0. \tag{3.6}$$

We can see from this that a decrease in pressure along a streamline leads to an increase in velocity, which fortunately makes intuitive sense: the fluid accelerates as it flows down a pressure gradient.

Note that this equation is found in a few different forms, for instance sometimes with specific enthalpy $h \equiv \varepsilon + P/\rho$, reducing the number of terms by one. In an ideal gas $\varepsilon$ and $P/\rho$ are related simply by $P/\rho = (\gamma - 1)\varepsilon$ where $\gamma$ is the ratio of specific heats, so that $h = \gamma\varepsilon = (P/\rho)\gamma/(\gamma - 1)$.

It is informative to compare this equation with (2.27) derived in section 2.4.3 for the case of an unsteady irrotational barotropic inviscid flow, as opposed to the steady non-barotropic inviscid flow in this chapter. In chapter 2, the quantity $u^2/2 + h + \Phi$ plus $\partial\phi/\partial t$ was constant in space but not in time; in this chapter, a fluid element retains this quantity $u^2/2 + h + \Phi$ as it moves around (i.e. it is constant along streamlines), which reflects its origin in energy conservation. Different fluid elements though will generally have different values of this quantity.

### 3.1.1 Applications of Bernoulli's equation

One nice phenomenon which can be easily understood with the help of this equation is the Venturi meter which measures the flow of air through a pipe: see figure 3.2. Another is the calculation of the flow of water out of a hole in a barrel—there is an exercise on this at the end of the chapter.

A nice experiment to do at home is to insert a ping-pong ball into the upwards stream of air from a hairdryer. The ball hovers in the stream and does not, as one might intuitively expect, fall out sideways. This can be explained qualitatively with Bernoulli's principle.

The interpretation of various phenomena is however not always as straightforward as it seems and it is often easier to go back to the momentum equation, considering the acceleration of, and the forces on, a fluid element. A good example of this is the aeroplane wing: it is often said that the pressure above the wing is lower

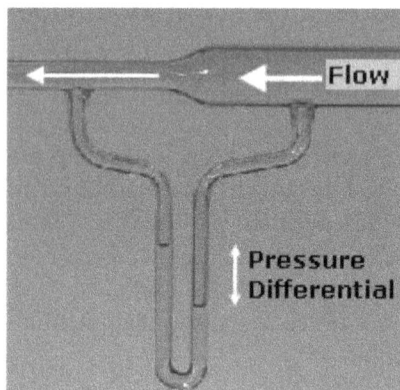

**Figure 3.2.** A Venturi meter, which measures the flow of air through a pipe. As the flow is constricted the velocity increases, meaning that $P/\rho$ must decrease. Since this is an adiabatic process we know that $P/\rho^\gamma$ is constant; given that $\gamma$ is always greater than unity we see that pressure must drop through the constriction. Note that the pressure and velocity difference change sign if the gas is moving supersonically through the pipe (see section 3.3), i.e. the velocity goes *down* as the fluid enters the constriction. (Reproduced from Wikimedia Commons under CC-BY-SA-3.0 by ComputerGeezer and Geof.)

than that below the wing because looking at the streamlines it is obvious that the air above the wing has further to travel and must therefore be moving faster, implying a lower pressure. This is misleading because there is no reason that the air flowing above the wing must meet up again with its former neighbour, so that it does not necessarily have to travel faster. To see the flawed argument, consider the lift generated by a *thin* curved aerofoil which is tilted with respect to the flow so that the air is deflected downwards: here the length of the streamlines above and below are the same, and yet the aerofoil still generates lift. A good example of this kind of aerofoil is the sail of a boat (in the situation when the boat is travelling upwind or perpendicular to the wind). How is the lift generated? The easiest explanation comes from consideration of the acceleration perpendicular to the streamlines: below the aerofoil the airflow must curve downwards. The only thing which can produce this acceleration is a pressure gradient perpendicular to the streamlines, namely such that the pressure near the aerofoil is higher than that further away. Above the aerofoil the fluid must also accelerate downwards so that the pressure just above the aerofoil must be lower than that further away. Since the pressure further away tends towards the ambient pressure, the pressure just above the wing must be lower than that just below it, accounting for the lift. Alternatively, one can think of the pressure changes as arising from the inertia of the oncoming fluid. Either way, Bernoulli's equation does not provide us with any quick explanation. See figure 3.3 for an illustration of the two types of aerofoil. We see that the lift comes essentially from the deflection of air downwards, which should have been obvious at the beginning from consideration of Newton's third law.

## 3.2 Subsonic and supersonic flow

In this section we look at properties of a steady compressible flow, finding quite different behaviour according to whether the flow is subsonic or supersonic[2].

Let us assume gravity is absent, so Bernoulli's equation is

$$h + \frac{1}{2}u^2 = h_0,$$

(3.7)

where $h \equiv \varepsilon + P/\rho$ is the specific enthalpy. The quantity $h_0$ is a constant along any given streamline and is equal to the enthalpy the fluid would have if the flow speed fell to zero. We are ignoring diffusive and heating and cooling processes in this chapter and so the entropy is constant along streamlines:

$$s = s_0.$$

(3.8)

Now, changes in enthalpy are given in (A.19) which in the case of *specific* quantities, i.e. per unit mass, becomes $dh = T\, ds + dP/\rho$. In an ideal fluid the term $T\, ds$ vanishes so we see that changes in $h$ and $P$ always have the same sign; this means

---

[2] When the author was 17 he was advised that 'supersonic flow separates the men from the boys' meaning, for those unfamiliar with the expression, that it is difficult. While it is true that it is non-intuitive—our intuition relates to everyday physics, which includes subsonic flow—it does not have to be difficult!

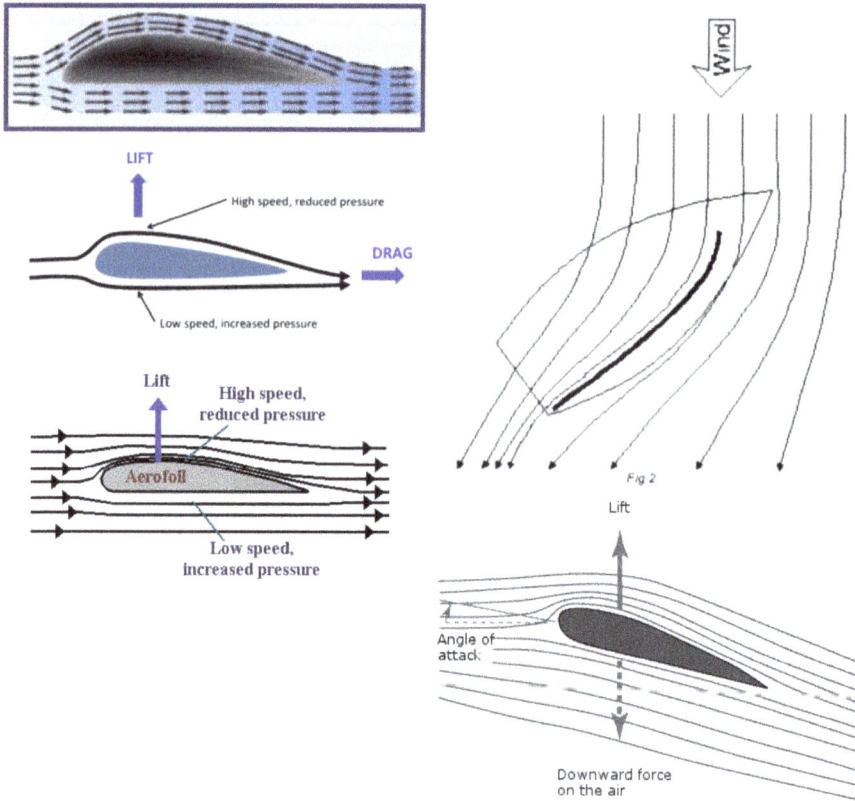

**Figure 3.3.** Left: examples of a commonly seen but misleading diagram of the Bernoulli effect producing lift on an aerofoil (the lower figure is from the NASA website[1]). The streamlines are longer over than under the aerofoil, resulting in a velocity and therefore pressure difference. However, it is in fact not at all obvious why the flow should be faster over the wing than under it. Upper right: the flow of air around the sail of a boat (https://byrdwords.wordpress.com/tag/sails/). This time, the streamlines immediately either side of the sail are of equal length, and yet we still have a velocity and pressure difference on either side. Lower right: a better diagram of wing lift (http://labman.phys.utk.edu/phys221core/modules/m8/turbulence.html).

that where the flow accelerates and the enthalpy drops—as is clear from (3.7)—the pressure must also drop. Physically this comes from the fact that the gas is being made to accelerate by the pressure gradient. While the maximum value of $h$ along a streamline is $h_0$, it is not immediately obvious what the maximum value of $u$ should be since we do not know the minimum value of $h$, which is a consequence of the fact that its absolute value has not yet been defined—as for internal energy $\varepsilon$, only changes $dh$ have been defined. The way out of this predicament is to recognise that the maximum velocity is where $P = 0$ since the flow cannot be accelerated any more if there is no pressure. We can then define the zero point of $h$ to be where temperature $T = 0$ and therefore $P = 0$. The maximum possible value of $u$ along

---

[1] The figure was downloaded from the NASA website in 2010. In the meantime NASA has replaced its own incorrect explanation of wing lift with a series of webpages debunking incorrect explanations of wing lift.

a streamline is then given by $u_{\max} = \sqrt{2h_0}$. This value is reached where the pressure goes to zero. A good example of this happening in nature is the solar wind: gas moves from a place with finite pressure to a place with essentially zero pressure and (ignoring gravity) the velocity is determined simply by the initial enthalpy. We have perfect conversion of thermal to kinetic energy, which if we think in terms of heat engines is only possible because the cold reservoir is at absolute zero. Microscopically we can think of a collection of particles with random thermal velocities being released into a vacuum where after some time a particle's position will depend only on its initial velocity and therefore its velocity is a function only of its position; there is no spread in velocities of particles in the same location and therefore the thermal energy has vanished.

The flow of fluid in terms of mass per unit area per unit time is $\rho \mathbf{u}$. This mass flux necessarily increases in the direction of the flow where streamlines converge and drops where they diverge. To calculate how it changes along a streamline we first look at the component of the momentum equation (2.1) along a streamline, which can be written

$$
\begin{aligned}
u \, \mathrm{d}u &= -\frac{\mathrm{d}P}{\rho} \\
&= -c^2 \frac{\mathrm{d}\rho}{\rho},
\end{aligned}
\tag{3.9}
$$

where the second line was arrived at using the relation $\mathrm{d}P = c^2 \, \mathrm{d}\rho$ where $c$ is the sound speed (see section 2.2), which is valid here because we are looking at adiabatic changes. We know from experience that $c$ is real, since we want $\mathrm{d}P$ and $\mathrm{d}\rho$ to have the same sign in an adiabatic expansion or compression. Returning to the mass flux $\rho \mathbf{u}$, we can now write its change along a streamline as

$$
\begin{aligned}
\mathrm{d}(\rho u) &= \rho \, \mathrm{d}u + u \, \mathrm{d}\rho \\
&= \rho \, \mathrm{d}u \left( 1 - \frac{u^2}{c^2} \right),
\end{aligned}
\tag{3.10}
$$

meaning that in a subsonic flow, converging streamlines accompany acceleration, whereas acceleration in a supersonic flow is found where the streamlines are diverging. The latter is outside of our everyday experience and therefore somewhat counterintuitive. In figure 3.4 we see this difference in terms of its effect on practical engineering[3].

Finally, we see from (3.10) that the maximum possible mass flux $\rho u$ along a given streamline must occur where $u = c$. This takes us onto the next section.

---

[3] If we think about a fluid as a collection of particles, there is an intuitive explanation of the increase in flow speed in the flared part of a rocket engine—the molecules hit the sides and rebound in a direction more closely parallel to the flow. Conversely if the tube gets narrower, the opposite happens and the molecules are reflected such that the component of the velocity parallel to the flow is decreased. While the kinetic energy of the molecule is unchanged, macroscopically we have converted flow kinetic energy into thermal energy.

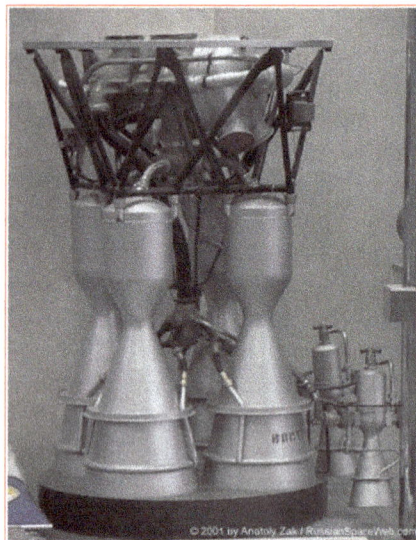

**Figure 3.4.** Left: water coming out of a hose. As children we learn to increase the flow speed by using a thumb to make the streamlines converge. (Image credit: Julija Sapic/Shutterstock.com). Right: rocket engines, like these from Vostok I, are flared to increase the exit speed. (Reproduced from Wikimedia Commons under CC-BY-SA-3.0 by Jud McCraine.)

## 3.3 Flow through a nozzle

Imagine we have a steady flow of compressible fluid through a tube of varying cross-section $A$ between two large volumes at pressures $P_0$ and $P_1$ where $P_0 > P_1$. We make the assumption that changes in the cross-section are gradual (i.e. that the diameter of the tube changes over length scales much larger than the diameter) and that the flow can be considered uniform across the cross-section of the tube. The fluid begins from rest in the first reservoir with enthalpy $h_0$ and entropy $s_0$, whose values in the tube are given by (3.7) and (3.8). Since the flow through the tube is steady the mass flux must be constant along the tube:

$$A\rho u = \text{const.} \tag{3.11}$$

The tube is connected smoothly to the first volume in such a way that the cross-section $A$ is very large where it joins the first reservoir and becomes smaller further away. The flow starts from rest and so is initially subsonic; we saw in section 3.2 that as the streamlines converge—in other words as $\rho u$ increases—the fluid accelerates.

If the pressure difference between the two reservoirs is small, the flow does not reach the sound speed and the pressure in the tube drops from $P_0$ at one end to $P_1$ at the other. It is important to note that the gas has already reached pressure $P_1$ as it exits the tube, giving lateral pressure balance between the emerging jet and the surroundings. The flow speed and therefore total flow in terms of mass per unit time can be calculated from (3.7) and (3.8).

The fractional pressure difference required to reach the sound speed can be calculated from the properties of the fluid. For instance, an ideal gas with a ratio of specific heats $\gamma$ has sound speed $c^2 = \gamma P/\rho$, enthalpy $h = c^2/(\gamma - 1)$, and the pressure and density during adiabatic expansion obey $P/\rho^\gamma = P_0/\rho_0^\gamma$. Using these relations and (3.7),

$$
\begin{aligned}
M^2 &= \frac{2}{c^2}(h_0 - h) \\
&= \frac{2}{\gamma - 1}\left(\frac{c_0^2}{c^2} - 1\right) \\
&= \frac{2}{\gamma - 1}\left(\left(\frac{P_0}{P}\right)^{\frac{\gamma-1}{\gamma}} - 1\right) \quad \Rightarrow \quad P_{M=1} = P_0\left(\frac{2}{\gamma + 1}\right)^{\frac{\gamma-1}{\gamma}}.
\end{aligned}
\tag{3.12}
$$

The numerical factor is 0.49 for $\gamma = 5/3$ and 0.53 for $\gamma = 7/5$. If the pressure in the second reservoir is equal to $P_{M=1}$ then the sound speed is reached exactly at the exit of the tube.

If the pressure of the second reservoir $P_1$ is less than $P_{M=1}$ then the behaviour depends on the shape of the tube. We can see from (3.10) and (3.11) that continued acceleration along the tube beyond the sound speed can only happen if $A$ increases. In other words, the tube becomes narrower further away from the first reservoir only until the sound speed is reached, and thereafter it must be flared. Such a tube is called a *de Laval nozzle* after the Swedish engineer; see figure 3.5. If it is not flared, then the fluid cannot accelerate beyond the sound speed and the pressure in the tube cannot drop below $P_{M=1}$, meaning that the remaining pressure drop from $P_{M=1}$ and $P_1$ must take place *after* the fluid has exited the tube, while the tube fluid is mixing into the ambient fluid. In a flared tube, however, the fluid can continue to accelerate past the *throat* or *sonic point*, driven by the remaining drop in pressure from $P_{M=1}$ downwards. The pressure after the sonic point can be calculated simply from the

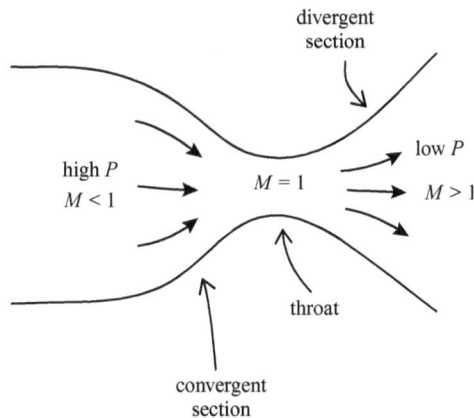

**Figure 3.5.** The de Laval nozzle.

cross-section $A$. If the cross-section at the exit is such that the pressure in the tube is greater than $P_1$, then the remaining pressure drop takes place outside of the tube as it does with an unflared tube when $P_1 < P_{M=1}$. If the pressure calculated at the exit is *lower* than $P_1$ then the flow has a tendency to break away from the boundaries of the tube and a stationary shock wave enters the tube, the details of which are very much an engineering problem and beyond the scope of this book.

## 3.4 Stellar winds and accretion

In this section we build on the analysis of nozzles to look at two very important astrophysical settings—accretion and stellar winds. Fortunately it turns out that both can be analysed with the same set of equations.

Let us imagine a steady, spherically-symmetric wind with velocity $u$ coming from a star. Negative $u$ signifies accretion. All quantities depend only on $r$, the distance from the origin. In nozzle parlance, the cross-section area of the flow is $4\pi r^2$ and so conservation of mass gives $4\pi r^2 \rho u =$ const, which we shall use later in the form:

$$-\frac{1}{\rho}\frac{d\rho}{dr} = \frac{1}{u}\frac{du}{dr} + \frac{1}{r^2}\frac{dr^2}{dr} = \frac{1}{2u^2}\frac{du^2}{dr} + \frac{2}{r}. \tag{3.13}$$

The momentum equation (in the radial direction) is

$$u\frac{du}{dr} = -\frac{1}{\rho}\frac{dP}{dr} - \frac{GM}{r^2}$$

$$\frac{1}{2}\frac{du^2}{dr} = -\frac{c^2}{\rho}\frac{d\rho}{dr} - \frac{1}{2}\frac{v_{esc}^2}{r}, \tag{3.14}$$

where $v_{esc}^2 = 2GM/r$ is the escape velocity, a function of radius, and where $dP = c^2\,d\rho$ was used as previously. This equation becomes, on substituting from (3.13),

$$\frac{1}{2}\frac{du^2}{dr} = c^2\left(\frac{1}{2u^2}\frac{du^2}{dr} + \frac{2}{r}\right) - \frac{1}{2}\frac{v_{esc}^2}{r}$$

$$\frac{1}{2}\frac{du^2}{dr}\left(1 - \frac{c^2}{u^2}\right) = \frac{1}{2r}\left(4c^2 - v_{esc}^2\right), \tag{3.15}$$

which is perhaps more elegantly written

$$\frac{r}{u^2}\frac{du^2}{dr} = \frac{4c^2 - v_{esc}^2}{u^2 - c^2}, \tag{3.16}$$

which we can also write in the form

$$\frac{r^2}{4r_s}\frac{1}{u^2}\frac{du^2}{dr} = \frac{1 - r/r_s}{1 - u^2/c^2} \qquad \text{where} \qquad r_s \equiv \frac{GM}{2c^2}, \tag{3.17}$$

where clearly, if the derivative of the speed is not to change sign somewhere then the denominator and nominator of the fraction on the right-hand side have to pass

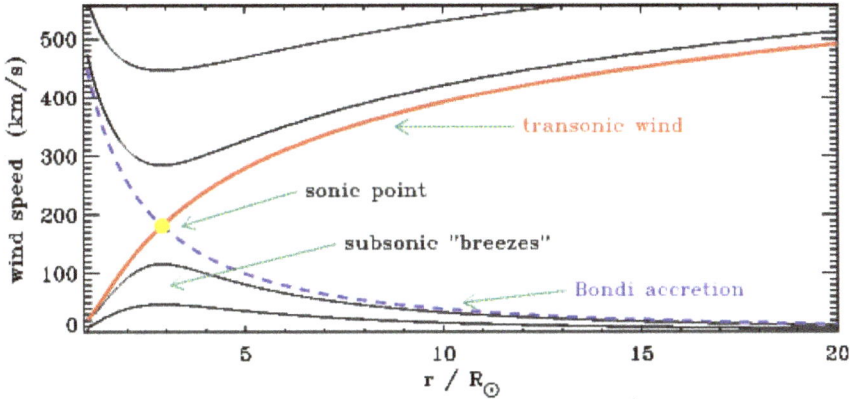

**Figure 3.6.** Solutions to (3.16). The interesting solutions are marked in red and blue: the transsonic wind and Bondi accretion. (Reproduced from http://lasp.colorado.edu/~cranmer/Old_Press/lambert_2004.html, with permission of Professor S R Cranmer, University of Colorado.)

through zero at the same place. Solutions to this equation are sketched in figure 3.6. They fall into six categories; two which pass through the sonic point at $r = r_s$ with $u = c$, and four in the quadrants separated by them. The main two solutions of interest are the two which pass through the sonic point; the solution which is supersonic at $r < r_s$ and subsonic at $r > r_s$ is accretion and the other is the stellar wind solution. This type of accretion and wind are normally attributed to Bondi (1952) and Parker (1965).

Because $r_s$ is a function of $c$ which is a function of $r$, the solutions in the adiabatic case are mathematically a bit complicated. Fortunately though it is possible to make a rough estimate of the rate of spherical accretion onto a star from an interstellar medium (ISM) of given temperature and density—we simply assume that the sound speed at the sonic point is not too different from the sound speed at infinity $c_0$, an assumption which is justified by the fact that the numerical factor found in (3.12) is not too different from unity (and because in astrophysics $2 \approx 1$). Taking the temperature of the ISM to be $10^4$ K, the sound speed is $c = \sqrt{\gamma R T / \mu_m} \approx 10$ km s$^{-1}$ if the gas is neutral hydrogen/helium. The accretion radius of a star of one solar mass is therefore $r_s = 7 \times 10^{13}$ cm, roughly the radius of Jupiter's orbit. As a star passes through the ISM the gas somewhat outside of this radius is affected little by the gravitational pull of the star, but inside this radius the gas is falling supersonically[4]. Using these numbers the accretion rate is

$$\dot{M} = 4\pi r_s^2 \rho c = \frac{\pi G^2 M^2 \rho}{c^3}, \tag{3.18}$$

which has the value $10^{11}$ g s$^{-1}$ or around $10^{-15} M_\odot$ yr$^{-1}$ if we take the ISM density to be 1 cm$^{-3}$, i.e. $\rho \sim 10^{-24}$ g cm$^{-3}$. Clearly this is not sufficient to form a star! This is

---

[4] This is the case if the star moves subsonically through the ISM. If the star is moving supersonically, the assumption of spherical symmetry must be dropped and Bondi accretion is replaced by *Bondi–Hoyle–Littleton accretion*.

why stars can only form in cold, dense environments; this state is achieved by radiative cooling. This, however, is only part of the solution; star formation is a tricky business.

This analysis explains how gas at rest accretes onto a star, but it does not explain the origin of stellar winds; for this it is necessary to look in more detail at the nature of a stellar atmosphere. We know that the Sun is surrounded by a hot tenuous medium called the *corona*, the Latin for 'crown'. We can attempt to find the structure of this atmosphere by assuming that the solar corona is static, and that the hydrostatic equation is satisfied:

$$\nabla P = \rho \mathbf{g} \quad \Rightarrow \quad \frac{dP}{dr} = -\rho g_r, \tag{3.19}$$

where the form on the right-hand side is given in spherical coordinates and spherical symmetry is assumed. In the static problem the energy equation reduces to

$$\nabla \cdot (K \nabla T) = 0 \quad \Rightarrow \quad \frac{d}{dr}\left(r^2 K \frac{dT}{dr}\right) = 0, \tag{3.20}$$

which comes from the theory of heat diffusion (see section 4.7), expressing the condition that the net heat flux into a fluid element is zero in a static equilibrium. It turns out that the thermal conductivity $K$ is proportional to $T^{5/2}$, which comes from kinetic theory of gases. Note the implicit assumption here that there are no heat sources or sinks, such as radiative losses. Imposing the boundary conditions $T = T_0$ at $r = r_0$, i.e. at the surface of the Sun, and $T = 0$ at infinity, the solution of the equation $r^2 T^{5/2} dT/dr = \text{const}$ is

$$\frac{T}{T_0} = \left(\frac{r}{r_0}\right)^{-2/7}. \tag{3.21}$$

Substituting this back into (3.19) gives

$$\frac{dP}{dr} = -\frac{r_0^2}{r^2} \frac{g_0 P}{R_\mu T} = -\left(\frac{r_0}{r}\right)^{12/7} \frac{g_0 P}{R_\mu T_0} = -\left(\frac{r_0}{r}\right)^{12/7} \frac{P}{H_0}, \tag{3.22}$$

where the equation of state $P = \rho R_\mu T$ and the inverse-square law $g = g_0 r_0^2/r^2$ have been used where $g_0$ is the gravitational acceleration at the solar surface. The pressure scale height at the surface $H_0 = R_\mu T_0/g_0$ has been defined. The solution is

$$\ln\left(\frac{P}{P_0}\right) = \frac{7 r_0}{5 H_0}\left[\left(\frac{r}{r_0}\right)^{5/7} - 1\right]. \tag{3.23}$$

The crux of the matter is that the pressure does not drop to zero as $r$ goes to infinity, in fact it drops only by around three orders of magnitude if, as is the case in reality, $H_0 \approx r_0/10$. This asymptotic pressure is much greater than the actual gas pressure in interplanetary space. Somewhere we have made an incorrect assumption! It turns out (not surprisingly) that the incorrect assumption is that the corona is in hydrostatic equilibrium: in fact, the material is moving outwards, accelerating as it does so and being observed as the *solar wind* as it passes by the Earth, where it has a velocity of around 400–600 km s$^{-1}$. Now, looking at (3.23) we can see that the extent to which the static solution is wrong, so to speak, depends on the ratio $H_0/r_0$. If the atmosphere of a star is very cold and this ratio is consequently very small, only a small correction to the static solution is required, i.e. the mass loss rate of the star is very small. This can be thought of as a situation where the sound velocity (comparable to the thermal velocity of the particles) is very much less than the escape velocity from the surface. To achieve significant mass loss, a star must have a sufficiently hot atmosphere so that the sound speed is not much less than the escape velocity. In this way the mass-loss rate of the Sun is around $10^{-14} M_{\odot}$ yr$^{-1}$, since the corona is heated to around a million kelvin by some process involving magnetic fields. (If the corona were hypothetically so hot that its sound speed approached and

**Figure 3.7.** A photo taken by the author in 1997 of the very bright comet Hale–Bopp. The white tail consists of dust and occupies the orbit of the comet. The blue tail consists of ions and points directly away from the Sun, in the opposite direction to the solar wind.

exceeded the escape velocity, the result would be explosive mass loss, corresponding to the everywhere-supersonic solution to the wind equation (3.16).)

This type of wind is called a *thermal wind* (Parker 1965). It is the dominant mechanism of mass loss in cool main-sequence stars with hot coronae such as the Sun. Higher-mass main-sequence stars lack a hot corona (a consequence of the lack of convection in the envelope) and so this mechanism does not work. In these stars winds can still be driven but by *radiative* mechanisms, especially as a star's luminosity approaches the Eddington limit. These winds are fundamentally different from thermal winds in that it is the outwards momentum imparted by photons on individual particles that drives the wind, rather than thermal energy deposited into the corona. Mathematically this takes the form of an extra term in the momentum equation (3.14). Incidentally, this question of whether mass loss is driven by momentum or energy from stars and supernovae is also often encountered in discussions of the loss of gas from star clusters, in particular from young clusters undergoing star formation.

Finally, note that the solar wind can be observed indirectly via its effect on comets—see figure 3.7.

## Exercises

### 3.1 Flow of water through a hole in a barrel
Consider a barrel containing water with a hole through which water is exiting. The hole has cross-sectional area $A$, which is small compared to the size of the barrel. Use Bernoulli's equation to calculate the time taken for a barrel containing volume of water $V$ and height $h$ to empty.

### 3.2 Hairdryer and ping-pong ball
Find a hairdrier and a ping-pong ball and suspend the latter in the upwards flow of air from the former (keeping the 'cold' button pressed to conserve energy!) Explain, with the help of Bernoulli's principle, how this works. How is the ball prevented from moving sideways out of the stream?

### 3.3 Lateral velocity
In figure 3.1 the streamtube is slightly twisted—there is apparently some vorticity and the fluid velocity is not quite parallel to the tube. However, in the derivation of Bernoulli's equation in section 3.1 the component of velocity perpendicular to the streamtube was completely ignored. Why was this possible?

### 3.4 Matching exit pressure to external pressure
We wish to construct a rocket engine which converts as much of the combustion thermal energy as possible into kinetic energy, in order to maximise propulsion. This means taking account of the external pressure. Comment on practical difficulties in building the optimum rocket engine to work in space.

### 3.5 Stellar winds and mass loss
An important but poorly understood process in stellar physics is mass loss. A very simple model of a stellar wind is the isothermal model. Assuming that the temperature is constant, calculate the mass-loss rate from a star as a function of its mass $M$, radius $R$, the temperature in the wind $T$ and the pressure at the base

of the wind (i.e. at $r = R$) $P$. Entering realistic numbers, estimate the mass-loss rate of the Sun, and comment.

**3.6 Accretion into galaxy clusters**

Galaxy clusters are the largest gravitationally-bound structures in the Universe. They grow by accreting matter from their surroundings. Estimate the accretion rate, given realistic parameters. Comment on the limits of assuming spherical symmetry.

# References

Parker E N 1965 Dynamical theory of the solar wind *Space Sci. Rev.* **4** 666–708

Bondi H 1952 On spherically symmetrical accretion *Mon. Not. R. Astron. Soc.* **112** 195

# Essential Fluid Dynamics for Scientists

**Jonathan Braithwaite**

# Chapter 4

# Viscosity

We now turn our attention to the form of the viscous term in the momentum equation, $\mathbf{F}_{\text{visc}}$. First we take a look at the equations and then use them to solve some simple problems.

## 4.1 The viscous stress tensor

First it is helpful to write the momentum equation in a more suitable form:

$$\frac{\partial}{\partial t}(\rho u_i) = -\frac{\partial T_{ij}}{\partial x_j}, \tag{4.1}$$

where $T_{ij}$ is the *momentum flux tensor*. It is easily demonstrated that it must always be possible to write the equation of motion in this form in situations without body forces such as gravity, by integrating over volume and using Gauss' theorem to express the right-hand side as a surface integral; the left-hand side then represents the rate of change of momentum of the volume and the right-hand side the forces acting on it at the boundaries. Also, note the similarity with the mass conservation equation. The momentum flux tensor is given by

$$T_{ij} = \rho u_i u_j + P\delta_{ij} - S_{ij}. \tag{4.2}$$

The term $\rho u_i u_j$ is often called the *Reynolds stress*, while the second and third terms together are called the *stress tensor*—where the viscous part thereof, $S_{ij}$, is called the *viscous stress tensor*.

In finding the form of $S_{ij}$, the following axioms must be adhered to. First, it must vanish in the case of a uniform velocity, which means that terms containing the velocity must be absent, and that it must instead be made up of velocity gradients. Secondly, we know that the viscous stress is linear in these gradients. So, the tensor must consist of only terms like $\partial u_i/\partial x_j$. Now, a non-zero value of $\partial u_i/\partial x_j - \partial u_j/\partial x_i$ represents a uniform rotation of the fluid, in which case the viscous stress must also

doi:10.1088/978-1-6817-4597-8ch4

vanish; this means that only terms $\partial u_i/\partial x_j + \partial u_j/\partial x_i$ are permissible, which represent a change of size or shape of the fluid elements. It is common at this juncture to introduce the *rate of strain tensor*

$$e_{ij} \equiv \frac{1}{2}\left(\frac{\partial u_i}{\partial x_j} + \frac{\partial u_j}{\partial x_i}\right). \tag{4.3}$$

There are further properties which follow from the symmetry and rotational symmetry between the three dimensions: the three diagonal elements must all have the same form; the tensor is symmetrical, i.e. $S_{ij} = S_{ji}$, and the off-diagonal elements must also have the same form. Therefore we can write

$$S_{ij} = ae_{ij} + b\delta_{ij}\frac{\partial u_k}{\partial x_k}, \tag{4.4}$$

where $a$ and $b$ are properties of the fluid in question. Now, it is observed in many fluids, including monatomic gases that—to a good approximation—no energy is dissipated during an *isotropic* compression or expansion. In other words, a fluid element can be compressed while its shape is preserved, then expanded again back to its original size, and the work done by the element during the expansion is equal to that done on the element during the compression. During such a change, $\partial u_1/\partial x_1 = \partial u_2/\partial x_2 = \partial u_3/\partial x_3$ and bearing this in mind we can rewrite (4.4) thus:

$$S_{ij} = \mu\left(2e_{ij} - \frac{2}{3}\delta_{ij}\frac{\partial u_k}{\partial x_k}\right) + \zeta\delta_{ij}\frac{\partial u_k}{\partial x_k}, \tag{4.5}$$

where $a$ and $b$ have been replaced by $\mu$ and $\zeta$ which are called the *shear* and *bulk viscosities*, respectively; in monatomic gases the bulk viscosity is zero. To find the physical reason for this, it is necessary to make a brief digression from 'classical' hydrodynamics and consider the constituent particles. We can consider the thermal energy in a gas in thermodynamic equilibrium as being divided equally (equipartition) between the various degrees of freedom, so that in a monatomic gas we have an energy per mole of $RT/2$ for the translational kinetic energy in each of the three dimensions and the total thermal energy per mole is $3RT/2$. During an isotropic expansion, energy is extracted at the same rate from kinetic energy in each of the three dimensions whereas the expansion of a gas in a cylinder–piston system extracts energy from just one dimension. In the latter case, the energies are brought out of equipartition and must gradually come back to equipartition; the finite time required to do this means that the pressure exerted on the piston during expansion is lower than it would be if the energy was redistributed instantly. During a compression the pressure on the piston is higher, therefore a net work must be done on the gas over a cycle consisting of expansion followed by compression; this work appears in the system as heat energy. This difference between the irreversibility of an isotropic and a non-isotropic change in volume is the origin of the second term inside the brackets in equation (4.5), ensuring that the stress tensor becomes zero in the isotropic expansion case. Now, a similar process occurs during the expansion of a gas made from diatomic or more complex molecules; at thermodynamic equilibrium, energy is split equally between not only the three translational kinetic energies but also the rotational kinetic energy (of which there are two degrees of freedom in the case of diatomic molecules such as those which make up the major fraction of the Earth's atmosphere). Even during an isotropic expansion, kinetic energy is extracted from the three translational degress of freedom but not from the rotational and the lag between the two gives rise to the same kind of dissipation as in the case of monatomic gas in a piston. This is the origin of bulk viscosity.

Recall that in section 1.1 we saw that the fluid approximation consists amongst other things in assuming that the mean-free path of particles is very much less than any other length scales of interest. This is because the idea of a local thermodynamic equilibrium is meaningful only in a fluid element at least as large as the mean-free path. Here, we have seen that the finite mean-free path, or rather the finite collision timescale, gives rise to a lag between energies and non-equipartition between different degrees of freedom. Therefore in some sense, the viscous terms in the fluid equations can be considered as first-order in the mean-free path.

In the case of an incompressible flow (i.e. the volume of each fluid element is not changing) with a velocity shear, the viscous stress acts to reduce the shear by transporting momentum across the fluid. Microscopically, this comes from individual particles transporting their momentum to another location where the mean velocity is different. Many undergraduate syllabuses include the calculation of the shear viscosity of a gas from consideration of momentum transport of particles in a shear flow. In applications where accuracy is not important (e.g. astrophysics) it is not important to know the detail of the calculation, but just that the dynamic viscosity of a gas is approximately equal to density × sound speed × mean-free path. In fact, this result can be obtained from a simple dimensional analysis. Remember that the sound speed is roughly equal to the thermal speed of the particles.

Back to the fluid picture: so far we have found that the viscous force per unit volume, i.e. the term $F_{\text{visc}}$ on the right-hand side of (1.3) can be written

$$F_{\text{visc},i} = \frac{\partial S_{ij}}{\partial x_j} = \frac{\partial}{\partial x_j}\left\{\mu\left(2e_{ij} - \frac{2}{3}\delta_{ij}\frac{\partial u_k}{\partial x_k}\right) + \zeta\delta_{ij}\frac{\partial u_k}{\partial x_k}\right\}. \tag{4.6}$$

This is a rather complicated expression, and we can get a better intuitive understanding of the physics if we make some simplifications. First of all, in an incompressible flow we can drop the terms with the velocity divergence. Next, we can assume that the dynamic viscosity $\mu$ is a constant and can therefore be brought outside of the divergence, giving

$$F_{\text{visc},i} = \mu\frac{\partial}{\partial x_j}(2e_{ij}) = \mu\frac{\partial}{\partial x_j}\left(\frac{\partial u_i}{\partial x_j} + \frac{\partial u_j}{\partial x_i}\right) = \mu\frac{\partial^2 u_i}{\partial x_j^2}, \tag{4.7}$$

where the zero-divergence of the velocity field was once again used to remove the second half of the rate of strain tensor. Defining the *kinematic viscosity* $\nu \equiv \mu/\rho$ we can write the momentum equation as

$$\frac{d\mathbf{u}}{dt} = -\frac{1}{\rho}\nabla P + \mathbf{g} + \nu\nabla^2\mathbf{u}. \tag{4.8}$$

In this form it is easier to understand the action of viscosity—essentially it acts to smooth out variations in the velocity field. Where there is a local minimum in $u_x$, for example, $\nabla^2 u_x$ is positive and so the viscosity brings about an increase in $u_x$. We can see from the first and last terms in (4.8) that this smoothing happens on a timescale

$$\tau_{\text{visc}} \sim \frac{L^2}{\nu}, \tag{4.9}$$

where $L$ is the characteristic length scale.

The extra viscous term in the momentum equation (4.8) fundamentally changes the nature of the equations. Without it, a problem can be entirely specified if the perpendicular component of the velocity is set to zero at the boundaries, as we shall see in section 4.5 when we calculate the inviscid flow past a sphere. However, with the viscous term, which contains a second-order derivative of the velocity, this boundary condition is not sufficient and something else is needed to properly constrain the solution. What is needed is that not just the perpendicular component but also the parallel part of the velocity must go to zero at the boundary. In fact we already know this from everyday experience, for instance when trying to blow dust off a flat surface —some layer of dust always remains. As we shall see below, what is happening is that there is a thin boundary layer of strong velocity shear next to the surface where the velocity goes from zero at the surface to its value in the external flow.

## 4.2 Viscous heating

Having calculated the stress tensor and therefore the effect of viscosity on the velocity field, we can now calculate the energy dissipated as heat and add a term to the energy equation. First, we take the dot product of velocity with the momentum equation

$$\rho \mathbf{u} \cdot \frac{\partial \mathbf{u}}{\partial t} = -\rho \mathbf{u} \cdot (\mathbf{u} \cdot \nabla)\mathbf{u} - \mathbf{u} \cdot \nabla P + \mathbf{u}_i \frac{\partial S_{ij}}{\partial x_j}, \tag{4.10}$$

which allows us to calculate the Eulerian rate of change of kinetic energy density:

$$
\begin{aligned}
\frac{\partial E_{\text{kin}}}{\partial t} = \frac{\partial}{\partial t}\left(\frac{1}{2}\rho u^2\right) &= \frac{u^2}{2}\frac{\partial \rho}{\partial t} + \rho \mathbf{u} \cdot \frac{\partial \mathbf{u}}{\partial t} \\
&= -\frac{u^2}{2}\nabla \cdot (\rho \mathbf{u}) - \rho \mathbf{u} \cdot (\mathbf{u} \cdot \nabla)\mathbf{u} - \mathbf{u} \cdot \nabla P + u_i \frac{\partial S_{ij}}{\partial x_j} \\
&= -\frac{u^2}{2}\nabla \cdot (\rho \mathbf{u}) - \rho \mathbf{u} \cdot \nabla\left(\frac{u^2}{2}\right) - \nabla \cdot (Pu) + P\nabla \cdot \mathbf{u} \\
&\quad + \frac{\partial}{\partial x_j}(u_i S_{ij}) - S_{ij}\frac{\partial u_i}{\partial x_j} \\
&= -\nabla \cdot \left(\rho \mathbf{u}\frac{u^2}{2} + Pu - uS\right) + P\nabla \cdot \mathbf{u} - S_{ij}\frac{\partial u_i}{\partial x_j}.
\end{aligned}
\tag{4.11}
$$

The term inside the bracket is an energy flux; integrating the equation over the volume of a fluid system one can convert this term to a surface integral of the flux over the boundary. The next term represents reversible conversion between thermal and kinetic energy; note that this term also appears in the energy equation (1.11), with the opposite sign. The last term is viscous conversion of kinetic into thermal energy. Taking the symmetry into account we can write this viscous heating per unit time per unit volume ($q$ is defined per unit mass) as

$$\rho q_{\text{visc}} = S_{ij}\, e_{ij}. \tag{4.12}$$

Note that the viscous heating is not equal to the local rate at which work is being done against viscosity $-u \cdot F_{\text{visc}}$, i.e. (minus) the last term in (4.10), which depends on the velocity rather than just on velocity gradients; part of this work is simply the transfer of momentum from one fluid element to its neighbours, and only part of it is actually dissipated. In addition, it is clear that the viscous heating must depend only on velocity gradients, and be positive; it is easily verified that $q_{\text{visc}}$ given in (4.12) satisfies both of these requirements as long as the viscosities $\mu$ and $\zeta$ are positive.

## 4.3 Examples of viscous flow

One much studied example of viscous flow is that of flow through a pipe. Here, we shall look briefly at a physically similar but geometrically simpler case, that of flow between two planes.

The two parallel planes separated by a distance $a$ are represented by $y = 0$ and $y = a$, and the flow is in the $x$ direction (the $x$-component of the velocity is $u$), so that there is no dependence on $z$. The $x$ and $y$ components of the momentum equation (4.8) are

$$0 = -\frac{1}{\rho}\frac{\partial P}{\partial x} + \nu\frac{\partial^2 u}{\partial y^2}, \tag{4.13}$$

$$0 = -\frac{1}{\rho}\frac{\partial P}{\partial y}, \tag{4.14}$$

where it is assumed that the flow is incompressible, that kinematic viscosity $\nu$ is constant and that a steady flow has been established between the two planes. Now, we see from the second of these that pressure depends only on $x$, and looking at the first equation we see that the pressure gradient must be constant. Integrating the first equation twice in the $y$ direction, we obtain

$$0 = -\frac{y^2}{2}\frac{\partial P}{\partial x} + \mu u + Ay + B. \tag{4.15}$$

To find the constants $A$ and $B$ we need to specify the boundary conditions, i.e. that velocity $u = 0$ at $y = 0$ and $y = a$ (as mentioned at the end of section 4.1, all components of the velocity must go to zero at the boundaries). If we assume that the boundary plates are stationary and therefore set the velocity at both boundaries to zero—which is called *plane Poiseuille flow*—then we find that $A = (a/2)\mathrm{d}P/\mathrm{d}x$ and $B = 0$, so that

$$u = -\frac{a^2}{2\mu}\frac{\mathrm{d}P}{\mathrm{d}x}\left(\frac{y}{a} - \frac{y^2}{a^2}\right), \tag{4.16}$$

so the velocity is positive if the pressure gradient is negative, an intuitively understandable result: the fluid flows down the pressure gradient.

It is informative to briefly mention the case of *plane Couette flow*, which is like plane Poiseuille flow except that the pressure gradient is zero and the planes are moving relative to one another. Taking the upper plate to be moving with velocity $V$, the solution is

$$u = \frac{Vy}{a}. \tag{4.17}$$

The two flows are illustrated in figure 4.1.

The equations of viscous flow can be solved in some other, more complex situations. The first case above can be extended to the case of flow through a pipe of circular cross-section, in which case it is observed that if the flow speed is high and/or the viscosity is low, some instability sets in and there is a transition to turbulence.

A popular case for investigation in astrophysics is the extension of the second case to flow between two concentric cylinders, which is called *circular Couette flow* or sometimes just *Couette flow*. In this situation, interesting effects can be seen at high rotation rates and/or low viscosity, in which case the flow is often called *Taylor–Couette flow*—another example of regime change. This kind of 'regime change' is common in viscous flow, and is explored in the next two sections.

## 4.4 Similarity and dimensionless parameters

The value of $\nu$ can have an important effect on the properties of the flow. It is informative to compare the size of the viscous term in the momentum equation with the other terms, along the lines of (2.14), in a steady flow:

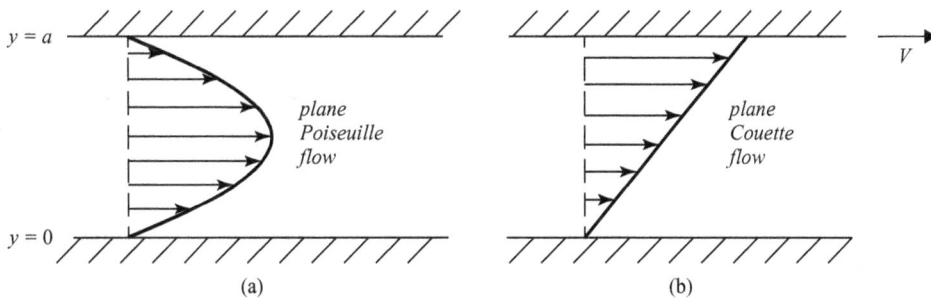

**Figure 4.1.** (a) Poiseuille and (b) Couette flow.

$$(\mathbf{u} \cdot \nabla)\mathbf{u} = -\frac{1}{\rho}\nabla P + \nu\nabla^2\mathbf{u}$$

$$\frac{U^2}{L} \qquad \frac{\delta P}{\rho L} \qquad \frac{\nu U}{L^2},$$

$$U^2 \qquad \frac{\delta P}{\rho} \qquad \frac{U^2}{\mathrm{Re}} \qquad \text{where } \mathrm{Re} \equiv \frac{UL}{\nu}, \tag{4.18}$$

where Re is called the *Reynolds number*. In the case of high Reynolds number the viscous term is small and the other two balance each other; conversely in the low Reynolds number case the viscous force balances the pressure gradient. Not surprisingly flows in these two regimes $\mathrm{Re} \ll 1$ and $\mathrm{Re} \gg 1$ have rather different properties which we explore in the following sections. Note however that a high Reynolds number does not generally mean that we can ignore viscous effects entirely; in fact as $\mathrm{Re} \rightarrow \infty$ the behaviour of a flow only tends towards the behaviour of a perfectly inviscid fluid with $\nu = 0$ in some special cases such as small oscillations; generally the presence of even a small viscosity has a fundamental effect, as we shall see in the next section.

This brings us to an important property of the equations of hydrodynamics, namely that since they contain no fundamental constants they are scalable. For instance, in the simplest case of a steady incompressible (subsonic) flow without gravity or viscosity, the nature of the flow is determined only by the geometry and not by the magnitudes of the various parameters, which are $L$, $U$ and $\rho$ (we can consider the pressure variation $\delta P$ as a function of these other parameters and so it cannot be set independently). We can set up two experiments with boundaries of the same geometry but with different densities, flow speeds and length scales, and the two flows will have identical geometry; the two flows are *similar*, hence the term *similarity flows*. This similarity is linked to the fact that it is impossible to make a dimensionless number out of combinations of $L$, $U$ and $\rho$. In compressible flow, the sound speed $c$ in the medium is an extra parameter and in order to make two similar flows with different $L$, $U$ and $\rho$, we also need them to have the same Mach number $M = U/c$ so that the fractional variations in density are the same. In the same way, similar *viscous* flows must have the same Reynolds number, which is the only dimensionless number it is possible to make from combinations of $L$, $U$ and $\rho$ and $\nu$ (except trivial functions of Re). This is obviously of enormous practical value when testing for instance the aerodynamics of boats in miniature water tanks. Another good example is that jets from stellar-mass black holes look very similar to those from 'supermassive' black holes eight orders of magnitude more massive.

The set of dimensionless parameters grows with every additional component. Similar flows with gravity must have the same *Froude number*, which is the ratio of the inertia to gravity, and similar *unsteady* flows must have the same value of the *Strouhal number*, the ratio of the $(u \cdot \nabla)u$ and $\partial u/\partial t$ terms. In summary we have

| | | | |
|---|---|---|---|
| Mach number | $\dfrac{\text{velocity}}{\text{sound speed}}$ | $\text{M} \equiv \dfrac{U}{c}$ | (4.19) |
| Reynolds number | $\dfrac{\text{inertia (steady)}}{\text{viscosity}}$ | $\text{Re} \equiv \dfrac{UL}{\nu}$ | (4.20) |
| Froude number | $\dfrac{\text{inertia (steady)}}{\text{gravity}}$ | $\text{Fr} \equiv \dfrac{U^2}{Lg}$ | (4.21) |
| Strouhal number | $\dfrac{\text{inertia (steady)}}{\text{inertia (unsteady)}}$ | $\text{St} \equiv \dfrac{UT}{L}.$ | (4.22) |

Note that the Strouhal number is infinity in a steady flow, approaches zero in small-amplitude waves (where it is a function of the other numbers if the oscillations are excited within the fluid rather than by some external agent), but is often of order unity in a variety of flows.

## 4.5 Regimes of viscous flow: example of flow past a solid body

We looked at the difference between flows with low and high Mach numbers in sections 3.2 and 2.3; in this section we look at the effects of having low and high Reynolds numbers, i.e. high and low viscosity, using the context of a solid body moving through a fluid and examining the drag force on the body. First of all though, we look at the consequences of having no viscosity at all. Note that throughout this section, we assume the fluid is incompressible (equivalent to $M \ll 1$).

To calculate the drag force on a body, it is useful to go into the frame of reference in which the body is at rest because in this frame the flow is steady and we lose all $\partial/\partial t$ terms. Imagine a solid sphere or radius $a$ moving at velocity $v$ through an ideal ($\nu = 0$) incompressible fluid. Transferring to the inertial frame where the body is stationary, the fluid at a large distance from the sphere is irrotational, so the fluid must everywhere be irrotational, and we can express the velocity as the gradient of a scalar (see section 2.4.3). In addition, the fluid is incompressible so that the continuity equation (1.8) reduces to $\nabla \cdot \mathbf{u} = 0$ (see section 2.3.1), so that the velocity potential $\phi$ must satisfy Laplace's equation $\nabla^2 \phi = 0$. The solution of this equation is a boundary value problem. The velocity potential has to satisfy two boundary conditions—that the flow tends towards uniform at infinity and that the radial component of velocity is zero at the surface of the sphere—but we impose in this inviscid case no condition on the tangential velocity at the surface of the sphere:

$$\phi \to -vr \cos \theta \quad \text{as } r \to \infty \quad \text{and} \quad \frac{\partial \phi}{\partial r} = 0 \quad \text{at } r = a, \tag{4.23}$$

using spherical coordinates comoving with the body where $r$ is the distance from the centre of the sphere and $\theta$ is the angle between the radius line and the direction of oncoming fluid. From undergraduate courses in electrostatics for instance, we know that the solutions to the Laplace equation in spherical coordinates are

$$\phi = A + (Br + Cr^{-2})\cos\theta + (Dr^2 + Er^{-3})(3\cos^2\theta - 1) + \dots \qquad (4.24)$$

We can obviously ignore $A$ and it follows from the boundary condition at infinity that $D$ and all coefficients of higher positive powers of $r$ are zero. The radial derivative of $\phi$ at $r = a$ is

$$\left(\frac{\partial\phi}{\partial r}\right)_{r=a} = (B - 2Ca^{-3})\cos\theta - 3Ea^{-4}(3\cos^2\theta - 1) - \dots \qquad (4.25)$$

and since this must be zero for all $\theta$, $E$ and higher coefficients must vanish. We also see of course that $B = 2Ca^{-3}$, leaving us with

$$\phi = -v\left(r + \frac{a^3}{2r^2}\right)\cos\theta. \qquad (4.26)$$

The appearance of this velocity field is illustrated in figure 4.2.

To calculate the drag force on the sphere we need now to integrate $P\cos\theta$ over the surface. From Bernoulli's equation we have

$$\frac{1}{2}(\nabla\phi)^2 + \frac{P}{\rho} = \text{const.} \qquad (4.27)$$

We can immediately see from (4.26) that the flow speed is symmetrical upwind and downwind, meaning that the pressure must also be symmetrical. The consequence of this is that the drag force vanishes! In fact we could have arrived at this conclusion much more quickly: in the steady state it is easy to see that no work is done on the fluid because after it passes by the sphere it returns exactly to its original state. There is no mechanism in an inviscid fluid to convert kinetic energy to thermal energy.

If we take this solution for the velocity field and calculate the corresponding size of the viscous term in the momentum equation we find that for high Reynolds numbers this term should be negligible compared to the other terms, and yet we know from experience that the drag force remains very much non-zero even at very high Reynolds numbers ($>10^{12}$). The drag coefficient (defined below) as a function of Reynolds number is plotted in figure 4.3. This shows that the assumption of zero viscosity can be an incredibly bad approximation for a fluid with low viscosity. The reason for this lies at the boundary between fluid and solid: in a fluid with finite viscosity, the fluid velocity parallel to the surface of the sphere is constrained to go to zero at the boundary. There is no solution including this extra boundary condition in the irrotational potential flow picture (we speak of an 'overconstrained' problem), so we must accept that there is at least some region in which the flow becomes rotational, i.e. develops a non-zero vorticity. In the case of low viscosity, this occurs only in a thin boundary layer near the surface of the object as well as sometimes in a larger volume behind the object, depending on its geometry. An inviscid irrotational solution applies elsewhere, but this boundary layer makes all the difference to the

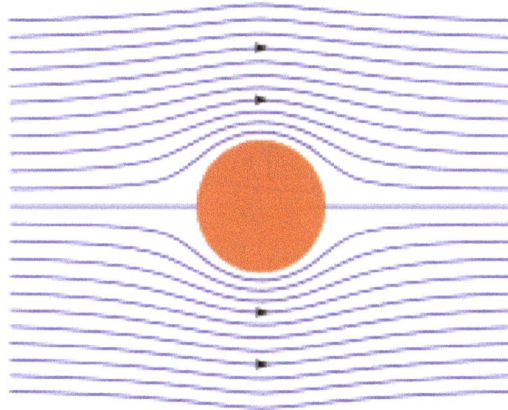

**Figure 4.2.** The flow of an inviscid incompressible fluid past a sphere. (Reproduced from Wikimedia Commons under CC-BY-SA-3.0 by Kraaiennest.)

drag force. On the right-hand side of figure 4.3 we see the difference flow regimes at different Reynolds numbers, and the boundary layers that form around the ball.

It is possible to calculate the drag force for very low Reynolds number. To do the complete calculation is tedious and if we just want astrophysical accuracy we can make do with a dimensional argument. Ignoring the inertial term in the momentum equation and equating the pressure gradient to the viscous term gives

$$\frac{1}{\rho}\nabla P = \nu \nabla^2 \mathbf{u} \qquad \Rightarrow \qquad F_{\mathrm{drag}} \approx L^2 \delta P \sim \rho \nu L U \qquad (4.28)$$

since the drag force can be thought of as the integration over the surface area $L^2$ of the body of the pressure variation. This type of flow is called the *Stokes regime*. The full calculation introduces just a numerical factor (in the case of a spherical body, a factor of $6\pi$ is introduced if $L$ is the radius of the sphere). We can now repeat the exercise for high Reynolds numbers:

$$\frac{1}{\rho}\nabla P = -(\mathbf{u} \cdot \nabla)\mathbf{u} \qquad \Rightarrow \qquad F_{\mathrm{drag}} \approx L^2 \delta P \sim \rho L^2 U^2. \qquad (4.29)$$

Here we are of course also missing a numerical factor. It is common in the literature to write the drag force as

$$F_{\mathrm{drag}} = \frac{1}{2} u^2 C_{\mathrm{d}} \rho A, \qquad (4.30)$$

where $u$ is the speed of the object through a stationary medium, $A$ is the cross-sectional area of the object as viewed from the direction of the oncoming fluid and $C_{\mathrm{d}}$ is a numerical factor (the 'drag coefficient') which depends on the geometry of the body as well as on the Reynolds number. The drag coefficient for a sphere is plotted in figure 4.3; the shape of this curve is a consequence of various phenomena to do

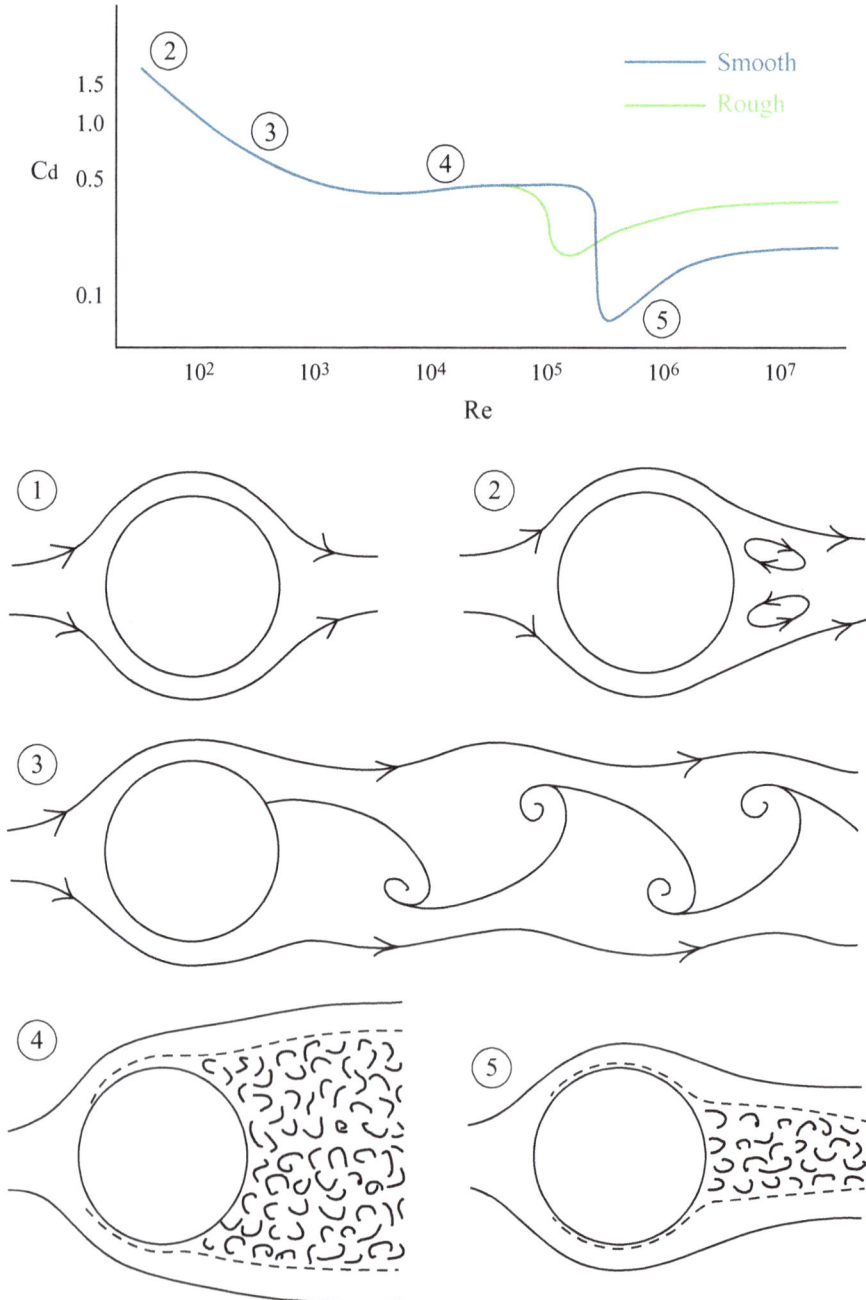

**Figure 4.3.** Top: drag coefficient of a sphere as a function of the Reynolds number. Bottom: flow past the ball at various Reynolds numbers. In regime 3 we see a *von Karmán vortex street* which, amongst other things, has been blamed (perhaps unjustly) for the collapse of at least one suspension bridge. In regimes 4 and 5 we see turbulence; the difference between the two is that in 4 the boundary layer is laminar, but in 5 it is turbulent. The transition between the two is influenced by the roughness of the sphere (which is why golf balls have dimples). Reproduced with permission from [2].

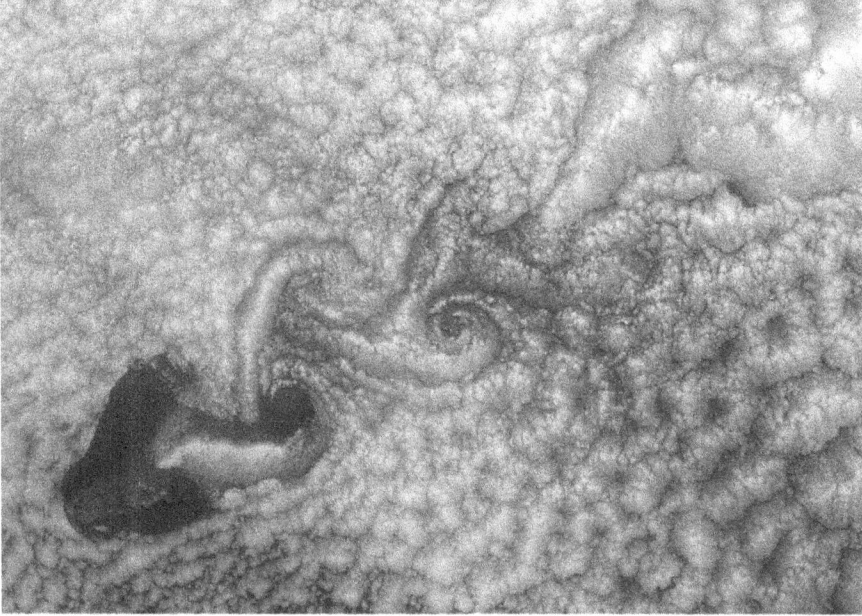

**Figure 4.4.** A von Karmán vortex street forms in the wake of an obstacle, in this case Guadalupe, off the west coast of Mexico. (Image credit: NASA/GSFC/JPL, MISR Team.)

with boundary layers and 'turbulence', which it is not necessary to explore here in detail[1]. Curiously, at intermediate Reynolds number a so-called *von Karmán vortex street* forms (figure 4.4).

Looking at the drag forces in the two regimes (4.28) and (4.29), the fundamental difference in their nature is apparent. We have a reasonable intuition for high-Re flow; even when we are moving around quite slowly, the air or water we have to move out of the way is flowing at very high Re. The situation is different if you are a very small creature such as a bacterium; basic activities like swimming work very differently at low and high Re—not having arms and legs is not a handicap at low Re because breaststroke doesn't work anyway. The reader is referred to an entertaining article on this topic by Purcell (1977).

## 4.6 Boundary layers

Boundary layers are important in many engineering as well as geophysical contexts and occasionally appear in astrophysics. We take a brief look at this very important phenomenon here.

Consider the flow around a thin plate aligned parallel to a uniform flow. In the inviscid case, the velocity field is completely unaffected by the presence of the plate;

---

[1] Comparing sizes of terms in the equations, one would expect a change in behaviour at a Reynolds number of about 1, so it is interesting to note that there are various regime changes at Reynolds numbers of 4 all the way up to $3 \times 10^5$. This illustrates something quite general, namely that factors of order unity are sometimes quite large (or small).

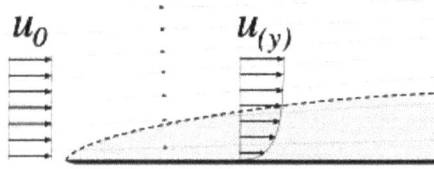

**Figure 4.5.** The boundary layer grows downstream. This image has been obtained by the author from the Wikimedia website where it was made available under a CC BY-SA 3.0 licence. It is included within this article on that basis. It is attributed to Flanker.

however, in a real fluid we must have $\mathbf{u} = \mathbf{0}$ at the surface of the plate. From experiments we know that the inviscid solution does still apply in the bulk of the volume but that there is a thin boundary layer where the viscous force is comparable to the other forces. This layer contains a strong velocity shear, which mean that vorticity is generated. The thickness of this boundary layer depends on the viscosity —it is thicker if the viscosity is greater. Furthermore, the layer grows downstream (figure 4.5), and it is possible to estimate this growth (see exercise below). Boundary layers may be either laminar or become turbulent if a shear instability develops (see section 5.3 for more about shear instability).

## 4.7 Heat diffusion

We know from everyday experience that heat flows from hot to cold. Without going too deeply into the thermodynamics, we can describe the flow of heat as

$$\mathbf{F}_{\text{heat}} = -K\nabla T, \tag{4.31}$$

where the heat flux has dimensions of energy per unit time per unit area and $K$ is the thermal conductivity. This equation contains the implicit assumption that the conductivity is isotropic, i.e. that the fluid conducts heat equally in all directions[2]. Now, to look at the effect of heat conduction on a body of fluid we need to know the net inflow/outflow of heat into/out of a fluid element; the net heat energy influx per unit volume per unit time $\rho q$ is given by

$$\rho q = \nabla \cdot (K\nabla T) \tag{4.32}$$

and if $K$ can be assumed constant throughout the fluid, we can simplify the spatial derivative to $K\nabla^2 T$. Furthermore we can define a thermal diffusivity

$$\chi \equiv \frac{K}{\rho c_p} \quad \text{and} \quad \text{Pr} \equiv \nu/\chi, \tag{4.33}$$

so that $\chi$ has the same units as viscosity $\nu$, defining also the *Prandtl number* Pr as the ratio of the two. In some situations, such as convection in stars, it is thought that a flow can behave quite differently according to whether the Prandtl number is greater than or less than unity.

---

[2] Isotropic conductivity is a valid assumption in most contexts of interest but in some cases, such as low-density plasmas where the mean-free path is greater than the gyration radius associated with the magnetic field present, the scalar $K$ must be replaced by a tensor.

# Exercises

### 4.1 Momentum equation

Verify that the momentum equation in Einstein summation notation (4.1) with (4.2) is equivalent to the vector-notation form (2.1), except for the gravity and viscous parts. Show that it is generally not possible to incorporate body forces, such as gravity, into the divergence-of-a-tensor form of (4.1).

### 4.2 Model testing

We are designing a boat which will sail at 4 m s$^{-1}$ and is 8 m long. We shall assume that the drag on the boat will be entirely due to buoyancy effects, i.e. generation of gravity waves. If we construct a model 50 cm long, at what speed should the water in the testing tank be moving past the model? (Hint: the Froude numbers must be the same.)

### 4.3 Growth of boundary layer

An infinitely thin solid sheet is inserted into a uniform flow such that sheet and flow are parallel. Argue that in the inviscid case, the flow is not affected. In the case of finite viscosity, show that a boundary layer forms and estimate the thickness of the boundary layer as a function of distance downstream. (Hint: look at the relative sizes of terms in the momentum equation.)

# References

[1]  Purcell E M 1977 Life at low Reynolds number *Am. J. Phys.* **45** 3–11
[2]  Brennan C E 1995 *Cavitation and Bubble Dynamics* (New York: Oxford University Press)

# Chapter 5

## Waves and instabilities

In this chapter we look at various types of waves and instabilities in various contexts. They can be categorised according to the nature of the restoring force. Sound waves, where the restoring force is the pressure gradient and which exist in any compressible fluid, we looked at in section 2.2. Here, we first look at various types of gravity wave, where the restoring force is gravity (obviously), and related instabilities which are driven by gravity and by a discontinuity in the velocity. At the end of this chapter we add self-gravity to the equation, giving rise to the so-called Jeans instability. In chapter 7 we shall look at waves which require rotation, such as inertial waves and Rossby waves, and in chapter 8 we look at magnetic waves.

## 5.1 Surface gravity waves

The first type of gravity wave to look at is the kind which propagates in a body of water. The water has uniform depth $h$ and the $z$ coordinate points upwards so the bottom and (equilibrium) surface of the water are at $z = -h$ and $z = 0$ respectively. Since we are not interested in sound waves, we can make life easier for ourselves by making the approximation that the water is incompressible. The first step is therefore to look at the equations of motion of an incompressible fluid in a gravitational field (see section 2.3.1); the continuity and momentum equations are then

$$\nabla \cdot \mathbf{u} = 0; \tag{5.1}$$

$$\frac{\partial \mathbf{u}}{\partial t} + (\mathbf{u} \cdot \nabla)\mathbf{u} = -\frac{1}{\rho}\nabla P - g\hat{\mathbf{z}}. \tag{5.2}$$

Sound waves having been filtered out by our assumption of incompressibility, the only possible restoring force for a wave in this system is gravity. Linearising the momentum equation we have

$$\frac{\partial \mathbf{u}}{\partial t} = -\frac{1}{\rho}\nabla P - g\hat{\mathbf{z}} \qquad (5.3)$$

in addition to

$$\nabla^2 P = 0, \qquad (5.4)$$

where the linearisation is valid as long as the amplitude of the wave is much less than both the wavelength $\lambda$ and the depth of the liquid $h$. At rest, the surface of the liquid is at height $z = 0$ and the lower boundary is at height $z = -h$. The perturbation to the height of the surface is $\zeta$. We shall consider only two dimensions, the horizontal dimension being $x$; the horizontal and vertical components of velocity are $u$ and $w$. The boundary conditions of the system are

$$w = 0 \text{ at } z = -h, \qquad w = \frac{d\zeta}{dt} = \frac{\partial \zeta}{\partial t} + u\frac{\partial \zeta}{\partial x} \text{ at } z = \zeta$$

$$\text{and } P = 0 \text{ at } z = \zeta, \qquad (5.5)$$

where the first two are known as *kinematic boundary conditions* and the third as a *dynamic boundary condition*. Now, since the motion is irrotational (since it is barotropic and inviscid and begins with zero vorticity; see section 2.4.3) we may express the velocity field as the gradient of a scalar potential $\psi$ with the incompressibility condition as Laplace's equation $\nabla^2\psi = 0$. The kinetic boundary conditions can be expressed as

$$\frac{\partial \psi}{\partial z} = 0 \text{ at } z = -h \qquad \text{and} \qquad \frac{\partial \psi}{\partial z} = \frac{d\zeta}{dt} \approx \frac{\partial \zeta}{\partial t} \text{ at } z = 0, \qquad (5.6)$$

where some linearisation has been performed on the second condition: the last term is dropped because it is second-order in small quantities and the conditions can be approximated to apply at $z = 0$ rather than $z = \zeta$. We now make use of the form of Bernoulli's equation applicable in unsteady, irrotational flows (2.27):

$$\frac{\partial \psi}{\partial t} + \frac{1}{2}(u^2 + w^2) + \frac{P}{\rho} + gz = f(t). \qquad (5.7)$$

The second term is second-order and can be dropped, and the term on the right-hand side can be absorbed into $\partial\psi/\partial t$. Substituting from here for $P$ into the dynamic boundary condition in (5.5) gives

$$\frac{\partial \psi}{\partial t} + g\zeta = 0 \qquad \text{at} \qquad z = 0, \qquad (5.8)$$

where the same replacement of $z = \zeta$ by $z = 0$ has been made as before. We now consider solutions of the form

$$\zeta = \hat{\zeta}e^{i(kx-\omega t)} \qquad \text{and} \qquad \psi = \hat{\psi}\, Z(z)e^{i(kx-\omega t)}. \qquad (5.9)$$

Substituting this $\psi$ into the incompressibility condition $\nabla^2\psi = 0$ gives

$$-k^2 + \frac{1}{Z}\frac{d^2 Z}{dz^2} = 0 \qquad \Rightarrow \qquad Z(z) = e^{kz} + ae^{-kz}, \qquad (5.10)$$

where only one constant $a$ is needed because we can absorb the rest into $\hat{\psi}$. The value of $a$ we find from the first kinetic boundary condition of (5.6):

$$ke^{k(-h)} - ake^{-k(-h)} = 0 \qquad \Rightarrow \qquad a = e^{-2kh}$$
$$\Rightarrow \qquad Z(z) = e^{kz} + e^{-kz-2kh}.$$

(5.11)

Now the second kinetic boundary condition of (5.6) and the dynamic boundary condition (5.8) become

$$k\hat{\psi}(1 - e^{-2kh}) + i\omega\hat{\zeta} = 0,$$

(5.12)

$$-i\omega\hat{\psi}(1 + e^{-2kh}) + g\hat{\zeta} = 0.$$

(5.13)

Eliminating $\hat{\psi}$ and $\hat{\zeta}$ from these two simultaneous equations we finally arrive at the dispersion relation, i.e. the relation between $\omega$ and $k$:

$$\omega^2 = gk \tanh kh.$$

(5.14)

We see immediately from this relation that there are two regimes—one long wavelength and one short wavelength with dispersion relations

$$\omega^2 \approx gk^2h \text{ where } kh \ll 1 \text{ and } \omega^2 \approx gk \text{ where } kh \gg 1.$$

(5.15)

The phase speed $\omega/k$ and group speed $d\omega/dk$ are readily calculated in the two cases. The phase speeds are:

$$\frac{\omega}{k} \approx \sqrt{gh} \text{ where } kh \ll 1 \text{ and } \frac{\omega}{k} \approx \sqrt{\frac{g}{k}} \text{ where } kh \gg 1.$$

(5.16)

In the long wavelength case, the group speed is the same as the phase speed and neither depend on the wavelength. We say that the waves are *non-dispersive*, meaning that a wave train consisting of various wavelengths will stay intact as it propagates. In the short wavelength case, on the other hand, the group velocity is a factor of two smaller than the phase velocity and more importantly both phase and group velocity do depend on the wavelength. If we create some waves by making a localised non-sinusoidal disturbance for a finite period of time, the initially super-imposed waves of different wavelength will propagate outwards at different speeds and so the shape of the waves will not be preserved; in fact the waveforms will approach sinusoidality. We call this kind of wave *dispersive*. Note that sound waves (section 2.2) are non-dispersive; without this property thunder would sound quite different and there would be problems with sound quality at open-air concerts; it would probably be necessary for concert-goers to wear headphones.

We can now examine in more detail the structure of these waves in the two limits; the difference is illustrated in figure 5.1. In the long wavelength limit these waves are normally called shallow water waves (meaning that the water is shallow compared to the wavelength); tidal waves and tsunamis are good examples of these at different frequencies (an example is shown in figure 5.2). The function $Z(z)$ depends only weakly on $z$ and $dZ/dz \approx 2k^2(h + z)$. This derivative appears in the vertical velocity

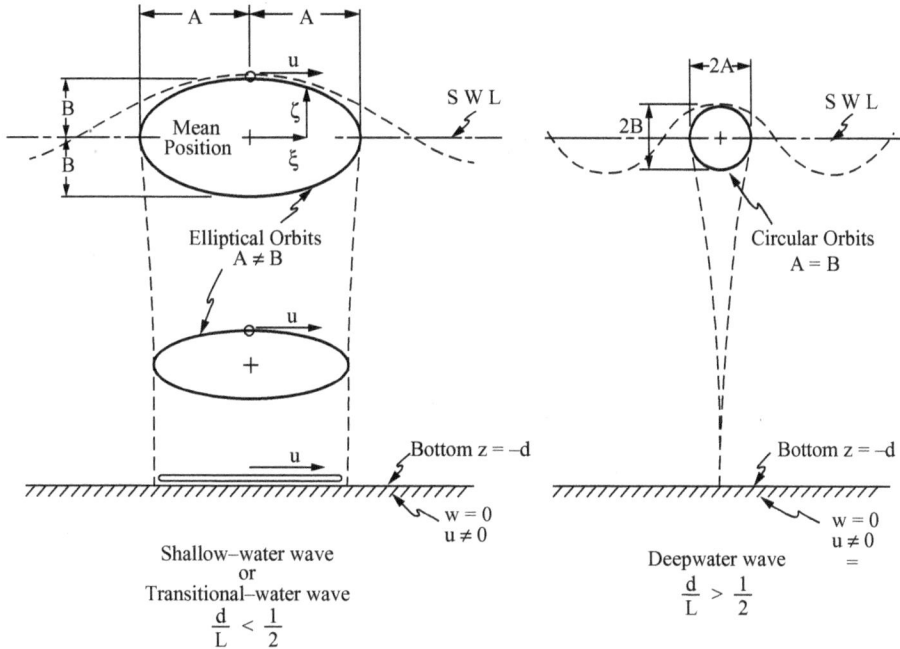

**Figure 5.1.** Particle paths in shallow and deep water waves. In shallow-water waves (meaning that the depth of the water is small compared to the wavelength) the motion goes right to the bottom; in the deep-water case, everything happens near the surface.

$w$ which varies therefore from a maximum on the surface linearly down to zero at $z = -h$. The horizontal velocity on the other hand is roughly constant with depth. In addition to this we see from (5.9) that $u$ and $w$ are 90° out of phase with each other, meaning that the particle paths are ellipses.

Deep water waves are excited on the sea by the wind (see section 5.3). Here, $Z(z)$ and therefore also the flow velocity decay exponentially away from the surface and are negligible at $z = -h$. The particle paths are circular, as can be seen by comparing the vertical and horizontal derivatives of $\psi$. Another example of a deep-water wave is shown in figure 5.3.

### 5.1.1 The shallow-water equations

We saw above that in the case where the characteristic length scale of motion in the horizontal direction is much greater than the depth of the water, the horizontal velocity of the fluid is independent of $z$. In this situation we can simplify the equations of motion at the beginning of the analysis; since $\partial u / \partial x$ is independent of height, and adding the $y$ dimension so is $\partial v / \partial y$, we see from (5.1) that $\partial w / \partial z$ must also be independent of height. This means that the continuity equation can be easily integrated over the depth of the water and thus be expressed as

$$\frac{\partial \zeta}{\partial t} = -h \nabla_h \cdot \mathbf{u}_h, \tag{5.17}$$

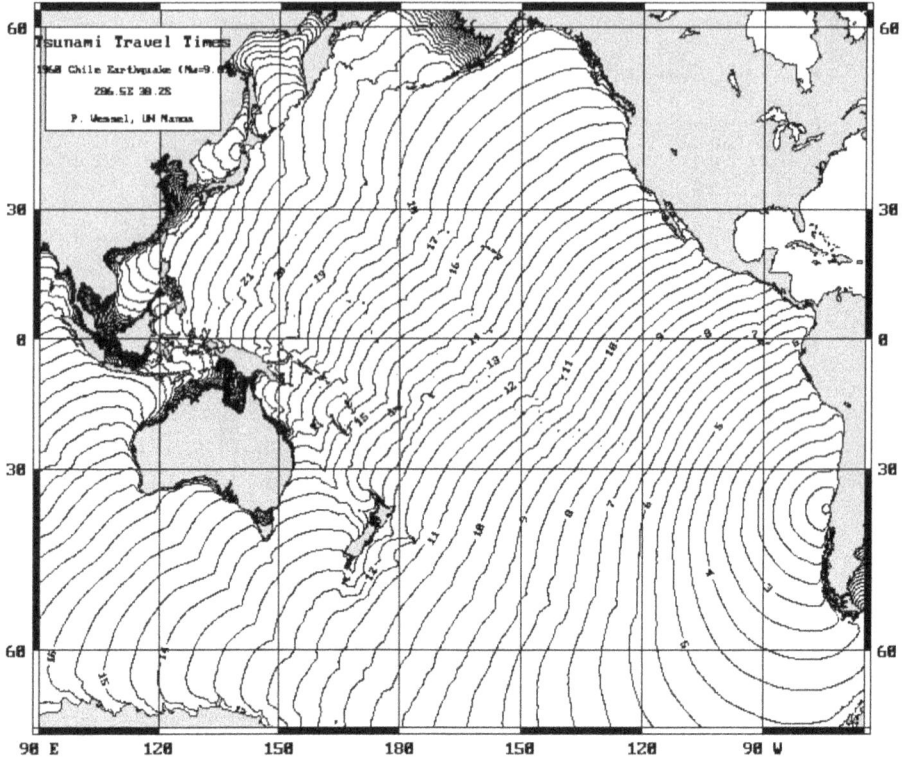

**Figure 5.2.** The propagation of the tsunami of 1960 across the Pacific. Shallow-water waves are non-dispersive, meaning that all wavelengths travel at the same speed. An earthquake might create something like a step function disturbance in the water, which can be considered as a Fourier superposition of all wavelengths. The short wavelengths are in the deep-water regime, propagate more slowly and are left behind; this smooths the original step function into a slope with a width comparable to the depth of the water. The long wavelengths however travel at the same speed so that this smoothed step retains its approximate form as it propagates across the ocean. Note finally how the speed changes in response to the changing depth and how this affects the direction of propagation. (Image credit: NOAA.)

where we have made the implicit assumption that $\zeta \ll h$, and where $\nabla_h$ and $\mathbf{u}_h$ are the horizontal components of the usual vectors $\nabla$ and $\mathbf{u}$. In addition the horizontal components of the momentum equation can be similarly integrated to give

$$\frac{\partial \mathbf{u}_h}{\partial t} + (\mathbf{u}_h \cdot \nabla_h)\mathbf{u}_h = -g\nabla_h \zeta, \tag{5.18}$$

which is a complete set of equations since the number of variables is only two: $\mathbf{u}_h$ and $\zeta$. Leaving out the $y$ direction and linearising to drop the second term on the left-hand side of (5.18) gives the same equations as we met with sound waves in section 2.2. The speed of the waves is $\sqrt{gh}$, which corresponds as expected to the long-wavelength limit in (5.16). We use the shallow water equations again in section 7.9 where a Coriolis acceleration is added.

**Figure 5.3.** Deep-water waves resulting from a disturbance (localised in both time and space) on the surface of a pond. Note that the longer wavelength modes have travelled further than the shorter wavelength modes—the wave is dispersive. (Image credit: 2xWilfinger/Shutterstock.com.)

## 5.2 One fluid on top of another fluid

We now turn our attention to the *Rayleigh–Taylor instability*, which occurs whenever a dense fluid lies in a gravitational field on top of a less dense fluid. The setup is similar to that in the previous section except that we now have two fluids with a boundary between them at $z = \zeta$. The fluid above has density $\rho_1$ and that below has $\rho_2$. Furthermore, to keep things simple we shall restrict ourselves to the case where both fluids are 'deep' compared to the length scale of the disturbance at the interface. As one might intuitively expect, the situation is unstable if $\rho_1 > \rho_2$. The kinetic boundary conditions (5.5) become

$$\text{As } z \to \infty: \qquad u_1 \to 0 \quad \text{and} \quad w_1 \to 0, \tag{5.19}$$

$$\text{As } z \to -\infty: \qquad u_2 \to 0 \quad \text{and} \quad w_2 \to 0, \tag{5.20}$$

$$\text{At } z = \zeta: \qquad w_1 = \frac{d\zeta}{dt} = \frac{\partial \zeta}{\partial t} + u_1 \frac{\partial \zeta}{\partial x} \quad \text{and} \quad w_2 = \frac{d\zeta}{dt} = \frac{\partial \zeta}{\partial t} + u_2 \frac{\partial \zeta}{\partial x}. \tag{5.21}$$

These become, using the velocity potentials $\psi_1$ and $\psi_2$ in the two fluids, and linearising (compare to (5.6)):

$$
\begin{aligned}
\nabla \psi_1 &\to \mathbf{0} && \text{as} && z \to \infty; \\
\nabla \psi_2 &\to \mathbf{0} && \text{as} && z \to -\infty; \\
\frac{\partial \psi_1}{\partial z} &= \frac{\partial \psi_2}{\partial z} = \frac{\partial \zeta}{\partial t} && \text{at} && z = 0.
\end{aligned}
\tag{5.22}
$$

The pressure boundary condition (5.5) becomes here $P_1 = P_2$ at $z = \zeta$, which via substitution of (2.27) and linearisation as before is

$$\rho_1 \left[ \frac{\partial \psi_1}{\partial t} + g\zeta \right] = \rho_2 \left[ \frac{\partial \psi_2}{\partial t} + g\zeta \right] \quad \text{at} \quad z = 0. \tag{5.23}$$

Assuming as before solutions of the form $\psi_1 = \hat{\psi}_1 \, Z(z) \exp\left[i(kx - \omega t)\right]$, this gives us the dispersion relation

$$\omega^2 = gk \frac{\rho_2 - \rho_1}{\rho_1 + \rho_2}. \tag{5.24}$$

Clearly, $\omega$ can be either real or imaginary, depending on which of the two fluids is more dense. If $\rho_1 < \rho_2$, $\omega$ is real then we have waves. If the two densities are very similar then the frequency can be quite low; this type of wave is seen in the ocean where there is a discontinuity in salinity. And by setting $\rho_1 = 0$ we arrive back at the

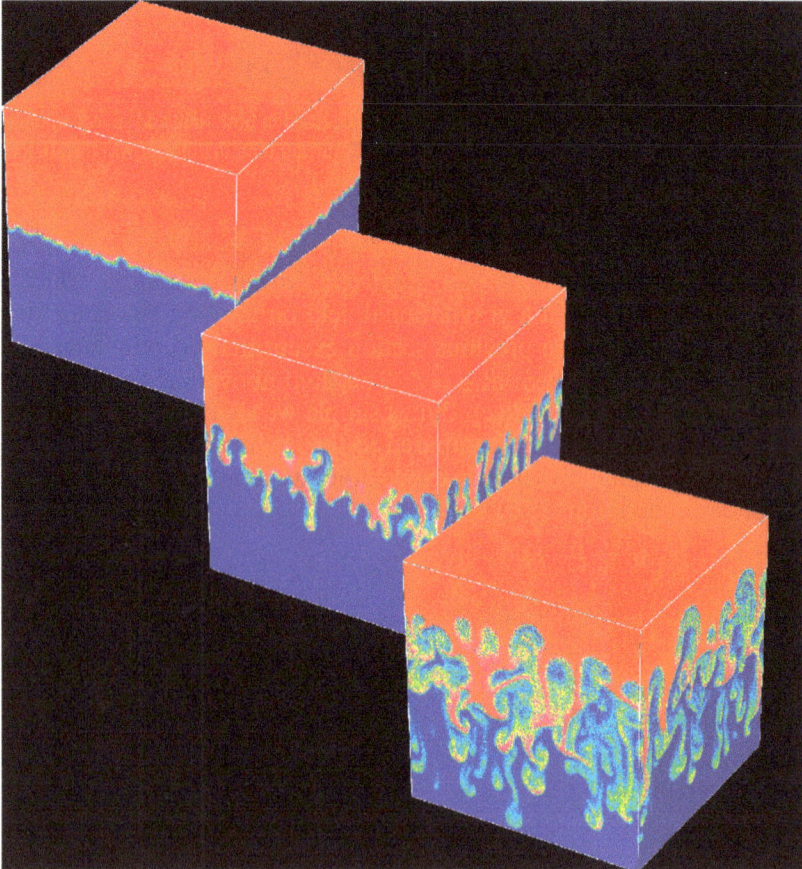

**Figure 5.4.** Simulations of the Rayleigh–Taylor instability in an incompressible fluid. Colour indicates density: red is dense and blue is less dense. Note that mixing occurs—the appearance of intermediate-density fluid (green). (Reproduced with permission of Dr K Kadau; see Kadau and Alder 2007.)

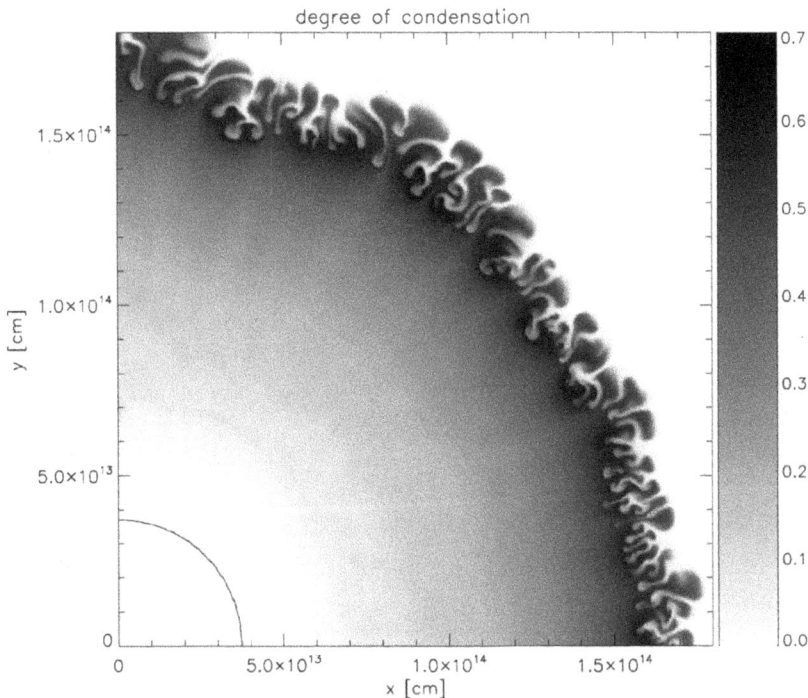

**Figure 5.5.** Simulations of the Rayleigh–Taylor instability in a stellar wind context. (Reproduced with permission from Woitke 2006.)

deep-water limit of the dispersion relation (5.15) we derived in the previous section. If $\rho_1 > \rho_2$ then $\omega$ is imaginary, meaning that the perturbation grows exponentially. The growth timescale in the unstable case is given by

$$\tau_{\text{R}-\text{T}} = \left( \frac{\rho_1 + \rho_2}{\rho_1 - \rho_2} \cdot \frac{\lambda/2\pi}{g} \right)^{1/2}. \tag{5.25}$$

This expression contains some ratio of the densities—in this case the difference between the densities is what drives the growth, so it is not surprising to see the fractional density difference in the expression for the timescale. The remaining part of this expression is simply the freefall timescale over a distance comparable to the wavelength; the shorter wavelengths grow more quickly. However, it is also important to note that the short wavelengths become non-linear at smaller amplitude $\zeta$ than the larger wavelengths, and that the larger wavelengths will ultimately be crucial for the eventual turnover and/or mixing of the fluids. Simulations of the Rayleigh–Taylor instability in two contexts are shown in figures 5.4 and 5.5.

## 5.3 Shear instability between two fluids

We now extend the analysis of the last two sections to include a background state in which the fluids have steady horizontal velocities and are flowing past one another.

**Figure 5.6.** Kelvin–Helmholtz instability in the laboratory. Two immiscible fluids of similar densities are placed in a long narrow tank and made to flow past one another by tipping the tank and letting gravity take its course. Image source: https://blogs.warwick.ac.uk/mjmarks/entry/two_entries_in.

The kinetic energy of this motion is a source of free energy to drive instability; the discontinuity becomes rippled, vortices appear and mixing takes place. This can be produced in the laboratory (figure 5.6). An obvious example of this instability in nature is the wind blowing over the surface of the sea, which excites the waves discussed in section 5.1, but many other examples occur in nature such as the boundary between an astrophysical jet and its surroundings. This instability at a surface of velocity discontinuity is called the *Kelvin–Helmholtz instability*[1].

Imagine two fluids separated by a horizontal planar discontinuity. The coordinate in the vertical direction is $z$ and that in the horizontal direction is $x$. The densities of the upper and lower fluids are $\rho_1$ and $\rho_2$, and the undisturbed velocities in the $x$ direction are $U_1$ and $U_2$. Note that there is no loss of generality, since it is always possible to change to a frame of reference in which both velocities are parallel to the $x$-axis. Making the further assumption of incompressibility and inviscidity we can express the velocity as the gradient of a scalar, since from Kelvin's circulation theorem we see that the vorticity must be zero everywhere at all times (see section 2.4.3). We can express the flow as the sum of the undisturbed flow and a (small) perturbation:

---

[1] Note that the astrophysics literature contains many references to Kelvin–Helmholtz instability in the case of a *continuous* shear flow such as those found in stars and discs, which is strictly speaking incorrect. The existence and nature of instability in such continuous shear flows is a topic of debate and is not covered here.

$$\phi_1 = U_1 x + \psi_1, \tag{5.26}$$

$$\phi_2 = U_2 x + \psi_2, \tag{5.27}$$

where $\phi$ and $\psi$ are the total and perturbation velocity potentials. The vertical perturbation to the position of the boundary is $\zeta$, and since the fluid does not pass through the boundary we have at the boundary

$$w_1 = (U_1 + u_1)\frac{\partial \zeta}{\partial x} + \frac{\partial \zeta}{\partial t}, \tag{5.28}$$

where $u_1$ and $w_1$ are the perturbation velocity components in the $x$ and $z$ directions. Linearising this equation, we can drop $u_1$ and take this condition to hold at $z = 0$ rather than $z = \zeta$. We have therefore at $z = 0$:

$$\frac{\partial \psi_1}{\partial z} = U_1 \frac{\partial \zeta}{\partial x} + \frac{\partial \zeta}{\partial t}, \tag{5.29}$$

$$\frac{\partial \psi_2}{\partial z} = U_2 \frac{\partial \zeta}{\partial x} + \frac{\partial \zeta}{\partial t}. \tag{5.30}$$

We also know that the pressure must be continuous across the boundary. We can find the pressure from the form of Bernoulli's equation applicable to unsteady irrotational flow (2.27):

$$\frac{\partial \phi}{\partial t} + \frac{1}{2}((U+u)^2 + w^2) + \frac{P}{\rho} + gz = f(t), \tag{5.31}$$

which applies on both sides of the discontinuity. Equating pressure on both sides gives

$$\rho_1 \left[ f_1(t) - \frac{\partial \phi_1}{\partial t} - \frac{1}{2}((U_1 + u_1)^2 + w_1^2) - g\zeta \right]$$
$$= \rho_2 \left[ f_2(t) - \frac{\partial \phi_2}{\partial t} - \frac{1}{2}((U_2 + u_2)^2 + w_2^2) - g\zeta \right]. \tag{5.32}$$

The unperturbed state must of course also satisfy this equation, so subtracting the unperturbed state and performing some reorganisation we find that at $z = 0$

$$\rho_1 \left[ \frac{\partial \psi_1}{\partial t} + U_1 \frac{\partial \psi_1}{\partial x} + g\zeta \right] = \rho_2 \left[ \frac{\partial \psi_2}{\partial t} + U_2 \frac{\partial \psi_2}{\partial x} + g\zeta \right], \tag{5.33}$$

where non-linear terms have been dropped. Since the motion at $z \to \pm\infty$ must vanish we see that the perturbations to $f_1$ and $f_2$ must also vanish since these quantities are constant in space; they can therefore be dropped from the equation above.

Let us assume a solution of the form $\psi_1 = \hat{\psi}_1 Z(z) \exp[i(kx - \omega t)]$ and similar form for $\psi_2$. For $\zeta$ there will be no dependence on $z$. Since we have assumed that the fluid is incompressible we have $\nabla^2 \psi = 0$, so that

$$\frac{Z''}{Z} - k^2 = 0 \qquad \Rightarrow \qquad Z = e^{\pm kz}, \tag{5.34}$$

where the sign in the exponent must be chosen so that solutions which diverge at infinity are dropped. Substituting these solutions into (5.29) and (5.30) gives

$$-k\hat{\psi}_1 = (-i\omega + U_1 ik)\hat{\zeta}, \tag{5.35}$$

$$k\hat{\psi}_2 = (-i\omega + U_2 ik)\hat{\zeta}. \tag{5.36}$$

We can also substitute the solutions into (5.33):

$$\rho_1(-i\omega + U_1 ik)\hat{\psi}_1 + \rho_1 g\hat{\zeta} = \rho_2(-i\omega + U_2 ik)\hat{\psi}_2 + \rho_2 g\hat{\zeta} \tag{5.37}$$

and using (5.35) and (5.36) to substitute for $\hat{\psi}_1$ and $\hat{\psi}_2$, this becomes

$$\rho_1(U_1 k - \omega)^2 + \rho_2(U_2 k - \omega)^2 = gk(\rho_2 - \rho_1). \tag{5.38}$$

The solutions are

$$\frac{\omega}{k} = \frac{\rho_1 U_1 + \rho_2 U_2 \pm \sqrt{(g/k)(\rho_2 - \rho_1)(\rho_2 + \rho_1) - \rho_1\rho_2(U_1 - U_2)^2}}{\rho_1 + \rho_2}. \tag{5.39}$$

It is a good idea to check at this stage that we recover (5.24) when we set $U_1 = U_2 = 0$. In the more general case, clearly there are no real roots to this quadratic equation (except for the trivial case $U_1 = U_2 = \omega/k$) without both non-zero gravity and $\rho_2 > \rho_1$; to be more precise the condition that the roots are real is

$$\frac{g}{k} > \frac{\rho_1\rho_2}{\rho_2^2 - \rho_1^2}(U_1 - U_2)^2. \tag{5.40}$$

If this condition is fulfilled, the waves are stable, i.e. their amplitude is constant. If not, the two solutions correspond to exponential growth and decay. Physically, one can think of stabilisation occurring if the energy released by the instability does not exceed the work which needs to be done against gravity to move the fluid vertically. An unstable mode will grow until non-linear effects become important; in the case of ocean waves excited by the wind, this happens at a particular ratio of amplitude to wavelength, and is visible as wave-breaking—at the onset of breaking, white froth appears. Note that the wavelength threshold increases with increasing velocity shear, explaining why the wavelength (and therefore height) of the largest waves on the sea is limited by the wind speed. Finally, a matter of terminology: if $\omega$ is real, the system is said to be stable; if it is imaginary, such as in the Rayleigh–Taylor case in the previous section where the displacement $\zeta$ increases exponentially, the system is unstable. If $\omega$ is complex and the solution represents oscillations of exponentially increasing amplitude, we say the system is *overstable*.

In many astrophysical contexts, gravity can be neglected, in which case a shear flow is unstable at all wavelengths. The timescale of the exponential growth is given by

$$\tau_{\text{K-H}} \equiv \frac{1}{\text{Im}(\omega)} = \frac{\rho_1 + \rho_2}{(\rho_1 \rho_2)^{1/2}} \frac{\lambda/2\pi}{\Delta U}, \tag{5.41}$$

where $\Delta U \equiv |U_1 - U_2|$ and $\lambda = 2\pi/k$. If the two densities are comparable to the growth time then ignoring factors of order unity we have $\tau_{\text{K-H}} \sim \lambda/\Delta U$, or in other words, it is equal to the time taken for one fluid to travel past the other a distance comparable to the wavelength. At shorter wavelength the instability grows faster, indeed the growth rate diverges at small wavelength and problems are prevented in reality by viscous damping of the shortest wavelengths.

We have now derived the dispersion relation and growth rate, but that has given us no insight into the physical mechanism which drives this instability. There are various ways to understand the mechanism; the simplest is to perhaps to think in terms of how we apply Bernoulli's equation to the flow around a solid body. Consider that one fluid is forced to flow over undulations in the discontinuity separating the two fluids, which requires its speed to increase; its pressure must therefore fall, encouraging the other fluid to flow further into the space occupied by the first fluid, thereby increasing the displacement $\zeta$.

The above describes the linear development of the instability. As the amplitude grows the instability becomes non-linear and eventually produces mixing of the two fluids. This provides an intuitive way of understanding where the instability comes from: if we mix the two fluids together until they move with a uniform velocity, while conserving linear momentum, the kinetic energy must be lower: imagine an observer in an inertial frame of reference such that $\rho_1 U_1 + \rho_2 U_2 = 0$, i.e. where the combined momentum of the two fluids is zero (or rather, where the combined momentum of layers of thickness $1/k$ on either side of the discontinuity is zero). Incidentally, note that the phase speed of the waves as given by the real part of $\omega/k$ in (5.39) vanishes in this frame. Now, initially the kinetic energy is obviously non-zero; however as the two fluids become mixed the kinetic energy does tend to zero. Every instability works off some kind of free energy; in this instance the free energy is the kinetic energy of the shear. This kinetic energy, which is directed in the $x$ direction, is converted first into kinetic energy in the $z$ direction and eventually also into thermal energy as viscosity damps motion on short length scales. In some sense the energy originally present wants to convert into other forms and ultimately into thermal, as that represents the greatest entropy.

If the two fluids are in fact the same, with $\rho_1 = \rho_2$, then the equations simplify somewhat. Gravity, if present, has no effect, and the shear discontinuity is unstable to all wavelengths. Consequently, if such 'vortex sheets' develop in any kind of flow, we should expect them to be unstable and break up.

Both the Kelvin–Helmholtz and the Rayleigh–Taylor instabilities have divergent growth rate as the wavelength goes to zero. In reality of course a fluid is viscous and without working through the equations properly we can estimate the shortest unstable wavelength if we know the kinematic viscosity $\nu$ (see chapter 4). The timescale on which any disturbance of wavenumber $k$ is damped is given by

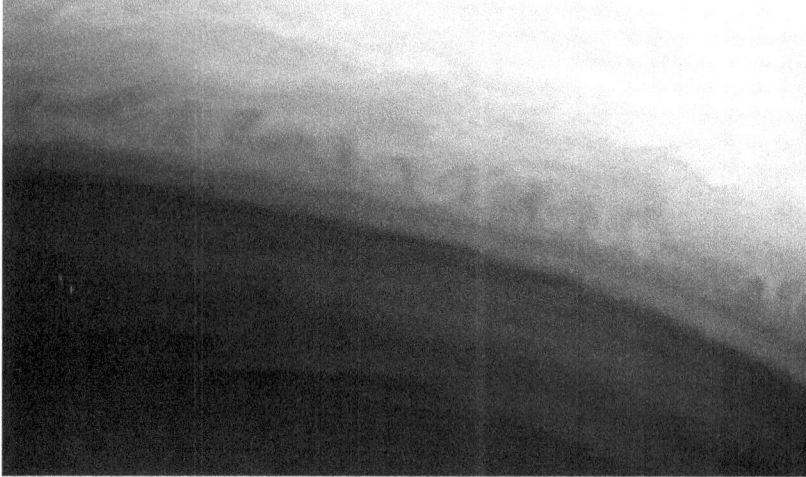

**Figure 5.7.** Photograph taken by the Cassini spacecraft of the Kelvin–Helmholtz instability in the atmosphere of Saturn. (Image credit: NASA/JPL/Space Science Institute.)

$$\tau_{\text{visc}} \approx \frac{1}{k^2 \nu} \tag{5.42}$$

and stabilisation occurs if this timescale is shorter than the growth timescale of the instability. Comparing with the instability growth times given by (5.41) and (5.25) it is easily verified that an upper limit on unstable wavenumber $k$ appears: short-wavelength modes are damped.

Some examples of the Kelvin–Helmholtz instability in nature are seen in figures 5.7, 5.8 and 5.9. Figure 5.10 is a snapshot from a simulation showing the instability.

## 5.4 Internal gravity waves

In section 5.1 we looked at gravity waves in a liquid with a free surface and generalised this in section 5.2 to waves between two fluids with a discontinuity in density; the restoring force comes from the tendency of gravity to flatten out this free surface. (We also saw that if the denser fluid is on top, it is unstable.) The assumption was made that each fluid has a constant density. In this section we look at fluids in which gravity waves can propagate without the need for any surface; for this reason, they are called *internal* gravity waves. Density variations within the fluid give rise to the restoring force as gravity pulls more strongly on the more dense fluid elements. Both (surface and internal) gravity waves are important in geophysics, while in astrophysics the internal waves are more important since astrophysics doesn't really have many sharp surfaces or indeed liquids[2].

---

[2] Note that in the astrophysical literature, these waves (or oscillations, if the waves are 'standing') are often referred to as 'g-modes', where the letter g refers to the nature of the restoring force. If on the other hand the restoring force is pressure, one speaks of 'p-modes' (i.e. sound waves), and of 'r-modes' if the restoring force comes from the rotation of the fluid body, via the Coriolis force (see chapter 7).

**Figure 5.8.** Shear flow in the atmosphere. The properties of water vapour near saturation enable us to 'see' the flow of air. This image has been obtained by the author(s) from the Wikimedia website where it was made available under a CC BY-SA 3.0 licence. It is included within this article on that basis. It is attributed to GRAHAMUK.

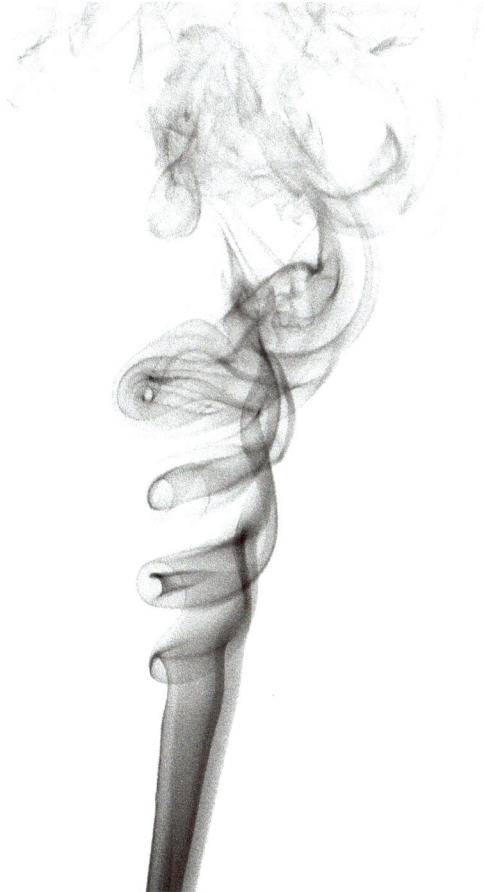

**Figure 5.9.** Rising cigarette smoke. A shear instability sets in and the flow becomes increasingly complicated. This behaviour is generic to jets of fluid passing through surrounding fluid at rest. (Image credit: Stanislav Spirin/Shutterstock.com.)

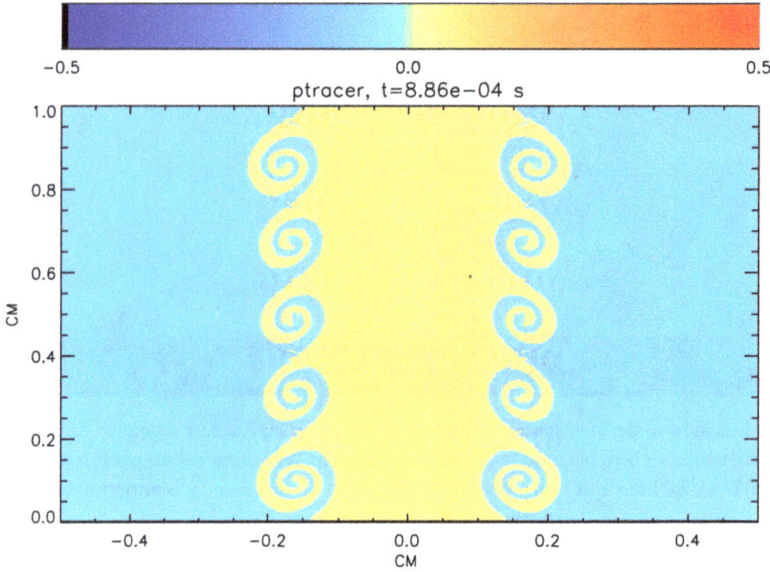

**Figure 5.10.** Kelvin–Helmholtz instability in a simulation. The flow is set up so that the fluid in the middle is moving upwards (in the figure) and the fluid on both sides is moving in the opposite direction; a perturbation is then added. The two fluids have a different value of an extra variable, the 'passive tracer', evolved such that its Langrangian time derivative is zero, making it possible to identify the two fluids (a technique common in numerical work). (Reproduced with permission from Braithwaite and Cavecchi 2012.)

Consider a fluid at rest in hydrostatic equilibrium, i.e. where $\partial P_0/\partial z = -\rho_0 g$ where the subscript 0 denotes the equilibrium quantities; gravity is directed downwards along the $z$ axis (see section 2.5.1). The equilibrium quantities are functions only of $z$. We are interested in small deviations from this equilibrium: the pressure field, for instance, becomes $P = P_0 + \delta P$. We want to look at subsonic motions, where the vertical length scale of interest is much less than the scale height $H_\rho$. Under these conditions, it makes sense to use the Boussinesq approximation (section 2.5.2). First we write down the equations: the momentum equation (2.32) including (2.39) to subsitute for $\delta\rho$, the continuity equation (2.35) and the energy equation (2.40):

$$\frac{\partial \mathbf{u}}{\partial t} + (\mathbf{u} \cdot \nabla)\mathbf{u} = -\frac{1}{\rho_0}\nabla \, \delta P + \frac{\alpha T g}{c_p} \, \hat{z} \, \delta s, \tag{5.43}$$

$$\nabla \cdot \mathbf{u} = 0, \tag{5.44}$$

$$\frac{\partial \, \delta s}{\partial t} + w\frac{\mathrm{d}s_0}{\mathrm{d}z} + w\frac{\partial \, \delta s}{\partial z} = \frac{q}{T}, \tag{5.45}$$

which are general equations in the Boussinesq approximation; the unit vector in the $z$ direction is $\hat{z}$. Recall that the thermal expansivity is defined as $\alpha = -(1/\rho)(\partial\rho/\partial T)_P$, equal to $1/T$ in an ideal gas. As previously in wave analysis, we linearise the equations. Accordingly we drop the second term on the left of (5.43) since it is second-order in $\mathbf{u}$, as well as the third term on the left-hand side of (5.45). We also set $q = 0$.

Assuming solutions of the form $\exp[i(\mathbf{k} \cdot \mathbf{r} - \omega t)]$, the three equations become

$$-i\omega\mathbf{u} = -\frac{1}{\rho_0}i\mathbf{k}\,\delta P + \frac{\alpha Tg}{c_p}\hat{\mathbf{z}}\,\delta s, \tag{5.46}$$

$$\mathbf{k} \cdot \mathbf{u} = 0, \tag{5.47}$$

$$-i\omega\,\delta s + w\frac{\mathrm{d}s_0}{\mathrm{d}z} = 0. \tag{5.48}$$

We immediately see from (5.47) that the motion is perpendicular to the wavevector $\mathbf{k}$. Combining these three equations to eliminate $\mathbf{u}$, $\delta P$ and $\delta s$ we find the dispersion relation

$$\omega^2 = N^2 \sin^2\theta, \tag{5.49}$$

where $\theta$ is the angle between the wavevector $\mathbf{k}$ and the vertical $\hat{\mathbf{z}}$, and $N$ is the *buoyancy frequency* or *Brunt–Väisälä frequency*, whose value is given by

$$N^2 = \frac{\alpha Tg}{c_p} \cdot \frac{\mathrm{d}s}{\mathrm{d}z}. \tag{5.50}$$

The first thing we see is that, since oscillations only occur if $\omega$ is real, we need a positive entropy gradient, assuming $\alpha$ is positive[3]. If this is not the case then instead of oscillations we get convective turnover, which is looked at in more detail in section 5.5. The second thing we see is that this dispersion relation is clearly quite different from those of surface gravity waves in that the frequency depends only on the angle of the wavevector and not on its magnitude. Physically this can be understood in the following way. A system hosting deep water surface gravity waves has no particular length scale; the only constant appearing in the equations is $g$. Therefore the waves can be made as small or as large as desired and correspondingly the frequency can have any value. In a continuously stratified fluid, however, the change in density takes place over a finite distance $H_\rho$. It is impossible to make the oscillations happen faster by reducing the length scale of the disturbance, because that also reduces the fractional difference in density and therefore the restoring force. In fact it is impossible to make the oscillations happen faster than the buoyancy frequency $N$.

Note also that the frequency goes to zero when the wavevector is vertical. In this case, the motion is entirely horizontal and it is clear that the restoring force vanishes. This is the reason that flows in stratified fluids tend to reside in horizontal surfaces; for instance the motion in the atmosphere is almost perfectly horizontal, except in those places where the entropy gradient becomes negative and convection appears.

---

[3] Exceptions to $\alpha > 0$ include water between 0 and 4 °C. This allows ice to form on the surface of a lake in the winter without all the water right down to the bottom necessarily having to cool to zero first.

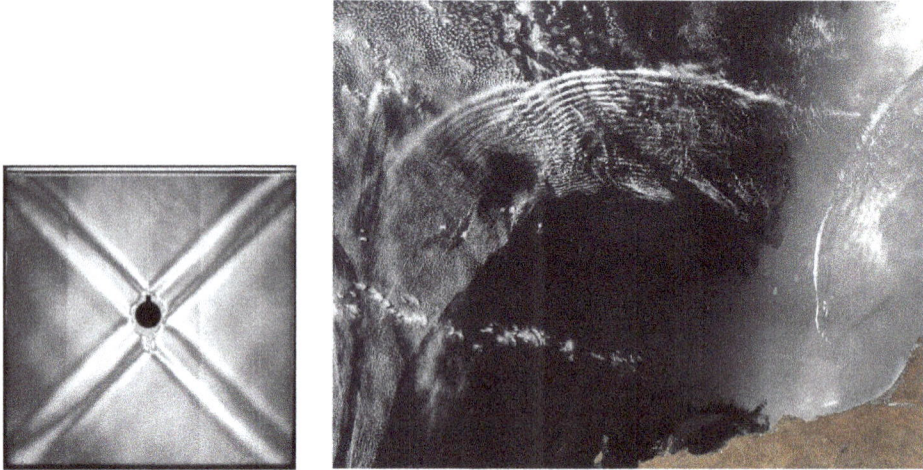

**Figure 5.11.** Left: internal gravity waves in the laboratory, propagating diagonally. (Reproduced with permission from Sakai *et al* 1997.) Right: atmospheric gravity waves off the coast of Australia.

The same is true of flows around stars; this manifests itself for instance in that chemical elements are mixed efficiently on surfaces of constant radius but that the mixing in radius is very slow in radiative (non-convective) zones.

## 5.5 Convection

We saw in the previous section that internal gravity waves can propagate in a gravitationally stratified fluid if the entropy gradient is positive, i.e. entropy increases upwards. If this is not the case, there is the tendency for an instability to appear which develops into something we normally call convection, with the result that heat energy is carried upwards. As with all instabilities though, this instability can be inhibited by diffusion and for historical reasons a particular case is discussed a lot in the literature.

*Rayleigh–Bénard convection* occurs in a fluid between upper and lower plates held at different temperatures and is readily investigated in the laboratory. A laminar (i.e. not turbulent) steady state can be obtained. Using the Boussinesq approximation (section 2.5.2) and looking at the sizes of terms in the equations (along the lines of section 4.4) it is apparent that there are only two degrees of freedom in the system, which can be expressed as the Prandtl number and the Rayleigh number defined thus:

$$\text{Pr} \equiv \frac{\nu}{\chi} \qquad \text{and} \qquad \text{Ra} \equiv \frac{\alpha g}{\nu \chi} \Delta T H^3, \tag{5.51}$$

where $\nu$ and $\chi$ are the kinematic viscosity and thermal diffusivity, $\alpha$ is the coefficient of thermal expansion, and $\Delta T$ is the temperature difference between the two plates, which are separated by a height $H$. It is observed that convection appears when Ra is above some critical value (which depends on Pr). Indeed a linear stability analysis predicts this. The definition of Ra is intuitively understandable: $\alpha$, $g$ and $\Delta T$ are

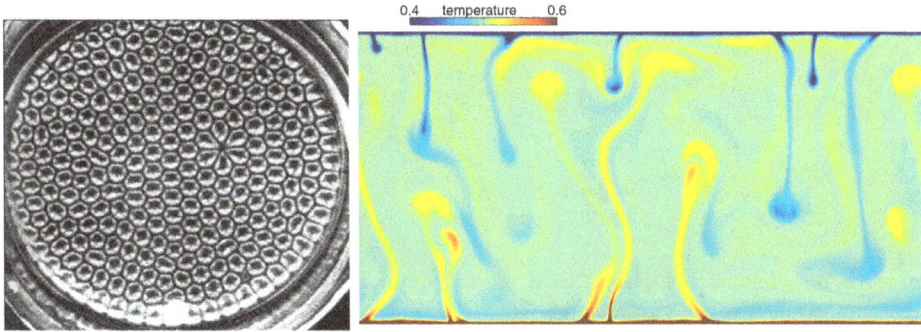

**Figure 5.12.** Rayleigh–Bénard convection. Left: at low Ra, either rolls or hexagonal cells appear. Right: a snapshot from a simulation (using the Boussinesq approximation) at high Ra: the convection becomes more complicated and time-dependent with boundary layers at top and bottom, which break off and enter the flow. (Reproduced with the kind permission of M Zimmermann, see Zimmermann 2009.)

driving the instability while diffusion of both kinds will inhibit it. $H$ has to be present to get the dimensions right.

Since Bénard's first experiments at the turn of the last century an enormous amount of work has been done on this kind of convection. It is an extremely popular topic for both physicists and mathematicians investigating the origin of order in systems. Some typical flow patterns are illustrated in figure 5.12. At low Ra the flow consists of large-scale, relatively steady convective cells, either in the form of rolls, or hexagonal in shape. At higher Rayleigh number the flow is more complex, fluctuates in time and takes on a random appearance. Experiments (laboratory and numerical) show that boundary layers form at top and bottom where the temperature gradient is high; these layers somehow peel away from the boundary and enter the main volume, where they are free to go across to the other side. This illustrates a general principle, namely that the boundary conditions can be important even far away from the boundaries.

In nature the Rayleigh number is often enormous, in which case the instability condition does not depend on diffusivities—the condition is simply that the entropy gradient is negative[4]. Another way to think about this stability condition is the following. Imagine a fluid package at a particular location, in pressure and thermal equilibrium with its surroundings at pressure $P$ and specific entropy $s$. Now let us displace this fluid element in the vertical direction to a place where the surroundings have pressure $P + \delta P$ and specific entropy $s + \delta s$. The package reaches pressure equilibrium with its new surroundings on the acoustic timescale, which is short compared to the other timescales as long as the motion is subsonic. However, the package retains its original entropy $s$ (see figure 5.13). If the displacement is upwards, i.e. if $\delta P < 0$, there is a restoring force downwards if the displaced package is denser than its new surroundings, which is the case if $\delta s > 0$ (assuming thermal

---

[4] In stellar physics this condition is known as the *Schwarzschild criterion*.

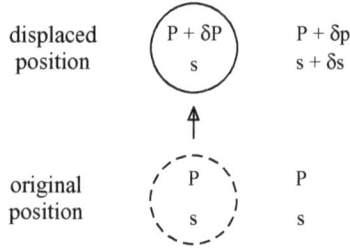

**Figure 5.13.** A fluid element is displaced from its original position and pressure and specific entropy $P$ and $s$ to a location where pressure and specific entropy are $P + \delta P$ and $s + \delta s$. The fluid element quickly comes into pressure equilibrium with its new surroundings, but retains its original specific entropy $s$.

expansivity $\alpha > 0$). Or if we displace downwards so that $\delta P > 0$, we need $\delta s < 0$ to have a restoring force downwards. Either way, for stability we need

$$\frac{\mathrm{d}s}{\mathrm{d}P} < 0. \tag{5.52}$$

It is informative to look at various examples of convection in nature. A good example is convection in the troposphere. Looking at cumulus clouds from the ground, it is not obvious what is going on. A good way to observe this convection is to get into a glider and be launched into it. Every glider is fitted with a 'vario' that measures the time derivative of pressure—roughly equivalent to the vertical component of the air's velocity[5]—and turns it into an audible signal of pitched beeps. Pilots notice that pockets of rising air can be relatively steady on the inside, often have abrupt boundaries and may go upwards a long way, much further than their horizontal extent.

Turning to space: on the Sun, we see convective cells which look a bit like those we see in the Rayleigh–Bénard case or in soup—see figure 5.14. They are of order 1000 km across, which is comparable to the scale height of the gas. They fluctuate over short timescales; indeed the lifetime of a convective cell on the surface of the Sun is comparable to the *turnover timescale* (simply the size of the cell divided by the flow speed), which is presumably the shortest timescale on which the flow could possibly change. At the solar photosphere there is a thermal instability that causes the temperature difference between updraft and downdraft to be suprisingly large; simulations show that the properties imparted onto the fluid at the photosphere are crucial in determining the nature of convection deep inside the Sun, with narrow downdrafts surviving a plunge of several scale heights (see e.g. Nordlund *et al* 2009 and references therein). Note that at the solar surface we have small kinematic viscosity $\nu$ but rather large thermal diffusivity $\chi$, since at the photosphere the mean-free-path of a photon is comparable to the scale height and so heat is transferred

---

[5] Albeit with a time lag $\Delta w/g$. This is normally only tenths of a second, much less than the time it takes to fly through the features in question.

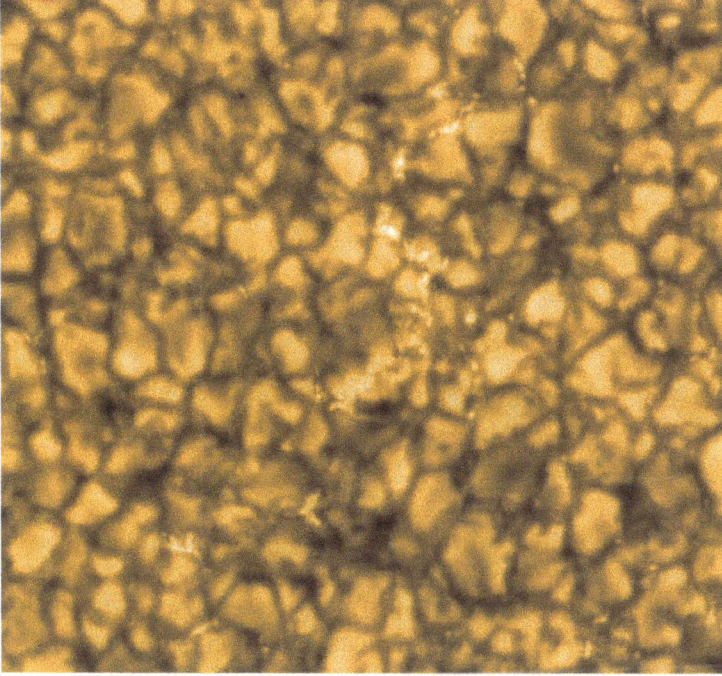

**Figure 5.14.** Convection in the Sun, as observed by the Hinode spacecraft. (Image credit: Hinode JAXA/ NASA/PPARC.)

rapidly. Deeper inside the Sun this is not the case and both diffusivities are small (in the sense that $UL/\nu$ and $UL/\chi$ are both much greater than unity).

One important feature that all forms of convection at high Rayleigh number have in common is high efficiency. The heat transported by convection feeds back into the background state until the entropy gradient is very small; in many contexts it is safe to assume constant entropy as soon as convection appears.

## 5.6 Baroclinic instability

In this section we take a brief look at a phenomenon related to convection. First, note that the stability condition (5.52) derived from our simple thought experiment is actually the stability condition for displacement *in a particular direction*. To have real stability, we require that this condition holds for *all* possible directions. This is the case if gradients $\nabla P$ and $\nabla s$ are exactly antiparallel, which requires the background state to have no horizontal variation. If $\nabla P$ and $\nabla s$ are merely *roughly* antiparallel, free energy can be released by displacing a fluid package approximately sideways in a direction such that it experiences a change in its surroundings with $\delta P < 0$ and $\delta s < 0$, in which case it has moved upwards and finds itself warmer than its new surroundings, and so experiences a buoyancy force pushing it further in the same direction (see figure 5.15). (A displacement in exactly the opposite direction is also unstable.) This is called the *baroclinic instability*, sometimes referred to as 'horizontal convection'. It is important in atmospheric physics and oceanography

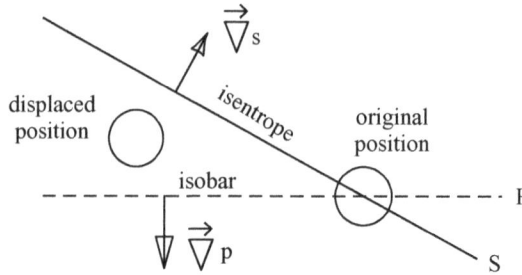

**Figure 5.15.** If $\nabla P$ and $\nabla_S$ are not exactly antiparallel, there is some range of directions of in which fluid elements can be displaced and release energy.

and probably also in rotating stars (in the radiative zones where normal convection is absent); see section 7.4.

## 5.7 Turbulence

Richard Feynmann described turbulence as the most important unsolved problem in classical physics. Werner Heisenberg apparently said 'When I meet God, I am going to ask him two questions: why relativity? And why turbulence? I really believe he will have an answer for the first.' A similar quotation is also ascribed to Horace Lamb. Sadly, in the decades since these gentlemen witticised on the topic, frustratingly little progress has been made.

The word 'turbulence' is not well defined and, in the absence of a theory, the phenomenon is purely observational. We see it in various flows with high Reynolds number; a turbulent flow looks complicated somehow, contains various length scales and fluctuates on various timescales. For instance, the air flow around a moving ball is laminar (i.e. not turbulent) at low speed and becomes turbulent above a certain speed, i.e. above a certain Reynolds number (see figure 4.3).

This observation has led to the popular misconception that turbulence always appears when the Reynolds number is sufficiently high. This obscures the origin of turbulence: namely that it represents the non-linear development of a fluid instability. At low Reynolds number an instability can be inhibited by viscosity; at high Reynolds number the instability is free to grow and become non-linear but it is crucial to note that *high Reynolds number alone will not result in turbulence—an instability is required.* There is often an obliging instability present—for instance in the air flowing around a ball we have a shear instability. Sometimes however there is no instability. For instance, it is now generally accepted that there is no purely hydrodynamical instability in an accretion disc and that consequently the flow should be laminar, despite an enormous Reynolds number (Ji *et al* 2006). (Accretion discs are however thought host a *magneto*hydrodynamical instability which allows them to live up to their name; see section 9.3.1.)

There is a popular theory to describe turbulence: that of Kolmogorov (1941). This model builds on the following assumptions:

1. that the observations imply the existence of a universal state of fluid flow at high Reynolds number that is independent of the particular conditions,

2. that this state has the property that it is local in space ('scale separation'),
3. that this state is also local in wavenumber space (the Richardson cell-inside-cell cascade).

An example: the convective zone of the Sun, accounting for the outer 2/7 of the radius—many scale heights deep. In this framework, the boundary conditions at top and bottom would only be relevant near those boundaries; in the bulk of the volume there would be some universal state, the same universal state as we would find in clouds or pipes or any other flow with a sufficiently high Reynolds number. In this picture, energy is injected at some large length scale and 'cascades' via 'eddies within eddies' down to smaller and smaller scales right down to a dissipative scale at which the Reynolds number is of order unity. A power spectrum can be derived that predicts the magnitude of these eddies as a function of length scale. For instance it is predicted that the energy depends on wavenumber as $E(k) \propto k^{-5/3}$. Various laboratory and numerical experiments have yielded power spectra with slopes asymptotically approaching $-5/3$ at sufficiently large $k$. However, other experiments have produced power spectra with exponents inconsistent with Kolmogorov's theory. Arguably there has been too much focus on power spectra. To measure such a relation from an experiment, a large amount of potentially interesting data is reduced to some function of wavenumber such as $E(k)$. In doing so, almost all of the data is thrown away; one can think in terms of converting the data into Fourier space and ignoring most of the magnitude information and all of the phase information. Indeed, many authors visualise their simulations *exclusively* in terms of power spectra. This is dangerous because a wide variety of qualitatively different flows can display identical power spectra. For instance, it is possible to construct a fluid flow containing just two length scales whose Fourier transform contains a continuous range of length scales, with a power spectrum $E(k)$ that looks identical to Kolmogorov's eddies-within-eddies flow with its continuous range of length scales. All in all, the predominance of the Fourier/cascade view of turbulence has resulted in an unproductive branch of fluid mechanics.

To gain an understanding of turbulence it is surely necessary to look at the flow in real space. One can observe turbulence in various contexts; various examples to do with convection were described in the previous section. Turbulence not involving convection can be observed for instance by standing on a bridge and watching the water pass by underneath. It pays to find a largeish river and to observe it some distance from the banks, or better still, from a boat. As long as the observation is not made too difficult by the presence of waves, it is often possible to make out chunks of water, perhaps 1–5 m across, which appear to move as a whole, rotating slowly and separated from each other by boundary regions with smaller detail.

In both laboratory and numerical experiments it has been established that vortex tubes are a general feature of turbulence. These vortex tubes account for a small fraction of the volume but a large fraction of the vorticity. The width of the tubes is a little more than the dissipation scale while their lengths are comparable to the macroscopic scale (see e.g. Vincent and Meneguzzi 1991). It appears that all the scales are correlated through these thin vortex tubes, whose lifetime is several times the large-scale turnover timescale.

**Figure 5.16.** A drawing from around 1510 by Leonardo da Vinci. He commented on the formation of vortices —an observation consistent with everyday experience but inconsistent with Kolmogorov's theory.

In a well-known turbulence experiment, fluid flows along a channel and through a grid. The flow is laminar upstream of the grid. Downstream we see asymptotic turbulence consisting of vortex tubes, in which transfer takes place coherently from the macroscopic scale to the smaller scales (see e.g. Mouri *et al* 2003).

The fact that a good theory has still eluded discovery is not due to lack of effort or intellectual strength. More likely it reflects the fact that the community has been led into a blind alley by Kolmogorov's theory. Even casual inspection of real flows shows that *all three* assumptions on which the theory is based, listed above, are demonstrably wrong. Assumption 1 is disproven by the existence of stable flows at high Re (as in the case of hydrodynamic accretion discs). Assumption 2 is contradicted by the fact that the interior of 'turbulent' flows is determined by the boundaries—boundary layers are advected through the flow; looking only locally at a volume in the interior of the flow, in ignorance of what is happening at the boundaries, this can alternatively be interpreted as 'intermittency effects'. Assumption 3 is contradicted by a look at the spatial structure of the flow, as seen in the aforementioned examples where the flow contains identifiable structures such as discontinuities and vortex tubes, rather than having the different wave-numbers (as seen in Fourier space) added together with random phases. Indeed, observational evidence ruling out this assumption dates back to the beginnings of modern physics itself—see figure 5.16.

## 5.8 The Jeans instability

In this chapter we have looked at various waves and instabilities where an important ingredient is gravity—a gravity that comes from an external source and appears in the equations in the form of constant field $\mathbf{g}$. In contrast, in this section we look at an instability that works with the gravity of the fluid itself, i.e. *self-gravity*. There are two ingredients in the restoring or driving force: pressure (as in sound waves) and self-gravity. In the momentum equation (2.1) the gravity term is given by $\mathbf{g} = -\nabla\Phi$ where $\Phi$ is the gravitational potential. For any body of density $\rho$, the gravitational potential $\Phi$ is given by the Poisson equation[6]

$$\nabla^2\Phi = 4\pi G\rho. \tag{5.53}$$

We can make the simplification (as in section 2.2) that all motion is in the $x$ direction and that all quantities have vanishing gradients in the $y$ and $z$ directions, making the momentum equation, continuity equation, and self-gravity equation:

$$\frac{\partial u}{\partial t} + u\frac{\partial u}{\partial x} = -\frac{1}{\rho}\frac{\partial P}{\partial x} - \frac{\partial\Phi}{\partial x}, \tag{5.54}$$

$$\frac{\partial\rho}{\partial t} = -\frac{\partial}{\partial x}(\rho u), \tag{5.55}$$

$$\frac{\partial^2\Phi}{\partial x^2} = 4\pi G\rho. \tag{5.56}$$

Just as in section 2.2, we linearise these equations. We first define some background equilibrium state with pressure $P_0$, density $\rho_0$ and zero velocity, writing pressure $P = P_0 + \delta P$, $\rho = \rho_0 + \delta\rho$, $\Phi = \Phi_0 + \delta\Phi$ where $\delta P \ll P_0$ and $\delta\rho \ll \rho_0$. It follows that the velocity $u$ is also small. As before, we define $c^2 \equiv \partial P/\partial\rho$. The linearised equations are

$$\frac{\partial u}{\partial t} = -\frac{1}{\rho_0}\frac{\partial\,\delta P}{\partial x} - \frac{\partial\,\delta\Phi}{\partial x}, \tag{5.57}$$

$$\frac{\partial\,\delta\rho}{\partial t} = -\rho_0\frac{\partial u}{\partial x}, \tag{5.58}$$

$$\frac{\partial^2\,\delta\Phi}{\partial x^2} = 4\pi G\,\delta\rho, \tag{5.59}$$

$$\delta P = c^2\,\delta\rho. \tag{5.60}$$

---

[6] The intermediate step here is $4\pi G\rho = -\nabla\cdot\mathbf{g}$. Note the similarity with Maxwell's equation $4\pi\rho_e = \nabla\cdot\mathbf{E}$. No constant is required in the electromagnetic case because it is built into the definition of the unit of charge (in c.g.s., but not SI units); the other difference is the minus sign.

Substituting from (5.60) for $\delta\rho$ into (5.58) and (5.59) gives

$$\frac{1}{c^2}\frac{\partial\,\delta P}{\partial t} = -\rho_0\frac{\partial u}{\partial x}, \tag{5.61}$$

$$\frac{\partial^2\,\delta\Phi}{\partial x^2} = \frac{4\pi G}{c^2}\,\delta P, \tag{5.62}$$

and then differentiating (5.57) w.r.t. $x$, substituting from (5.61) for $u$ and from (5.62) for $\Phi$, and tidying, gives

$$\frac{\partial^2\,\delta P}{\partial t^2} = c^2\frac{\partial^2\,\delta P}{\partial x^2} + 4\pi G\rho_0\,\delta P. \tag{5.63}$$

We now assume solutions of the form $\exp[i(kx - \omega t)]$, which upon substitution into (5.63) gives

$$\omega^2 = c^2k^2 - 4\pi G\rho_0, \tag{5.64}$$

which clearly gives oscillations (i.e. real $\omega$) only if the right-hand side is positive. The stability criterion is often expressed as an inequality in terms of the wavelength $\lambda = 2\pi/k$ as

$$\lambda < c\sqrt{\frac{\pi}{G\rho}}, \tag{5.65}$$

where the subscript on the density has been omitted. The critical wavelength of marginal stability is known as the *Jeans length*, named after the English astronomer James Jeans. Above this length scale, gas tends to collapse under its own gravity. Needless to say, this is pretty important in star formation.

## Exercises

### 5.1 Energy density and flux

(a) Show that (for the waves discussed here) the average potential and kinetic energies of a wave are equal. By integrating the work done per unit time and area $Pu$, derive the energy flux of the wave, and show that the group velocity is simply the ratio of energy flux to energy density.

(b) Explore what happens as a wave moves from deep water to ever shallower water, assuming that the energy flux is constant.

(c) By considering refraction, explain qualitatively what happens to the direction of propagation of a wave as it approaches a beach.

### 5.2 Waves at interface between two liquids

Construct a rudimentary experiment in a coffee cup to demonstrate the existence of waves at the interface between two fluids, the denser fluid below. (Hint: milk is denser than water.)

The excitation of waves between two layers of liquid is the explanation for a phenomenon called 'dead water' (død vann), where boats entering Norwegian fjords experience increased drag. The fjords contain fresh water lying on top of salty seawater.

### 5.3 Stokes' drift

By keeping second-order terms, show that in the deep water case the particle paths in a wave of finite amplitude are not quite circular but that averaged over a cycle a particle moves slightly in the direction of wave propagation.

### 5.4 Convection in soup

At home or during your next visit to a suitable eatery, try to observe convection in your soup. While all soups have roughly the same thermal expansivity $\alpha$ and thermal diffusivity $\chi$, there are significant variations in kinematic viscosity $\nu$; with the right soup you should be able to see convective cells. Miso soup is particularly suitable if it is sufficiently hot. Is there any fundamental difference between conditions in the soup and in the Rayleigh–Bénard case?

### 5.5 Thermal damping and double diffusive instability

In this exercise we consider displacing a fluid package from its equilibrium and examine the buoyancy force on it at its new location, with the goal of checking stability, as we saw in sections 5.5 and 5.6. Viscosity is ignored throughout.

(a) What happens in a stably-stratified fluid, i.e. where $ds/dz > 0$, when there is thermal diffusion? How is the buoyancy and consequently the oscillation of a fluid package affected by the diffusion of heat into and out of it while it is above and below its equilibrium position?

(b) Consider an ocean whose density deviation from some suitable average $\rho_0$ is given by

$$\frac{\delta\rho}{\rho_0} = -\alpha\,\delta T + \kappa\,\delta P + \beta\,\delta S, \tag{5.66}$$

where $S$ is the salt concentration. In addition, $\alpha$ and $\kappa$ are the thermal expansivity and isothermal compressibility (as usual) and $\beta$ is a coefficient describing the effect of salt on density; all three are considered constant here. This is simply an extension of (2.38). Assume that $(\partial T/\partial P)_s = 0$, so that the temperature of a fluid package is not affected by the change of pressure it experiences as it moves up and down adiabatically. Assume further that there is no diffusion of either heat or salt. Investigate the stability properties in terms of the vertical gradients of $T$ and $S$ and derive a stability condition.

(c) Looking at the condition you derived in part (b), in terms of stable/unstable and positive or negative vertical gradients of $T$ and $S$, there are six combinations. The three unstable combinations are dynamically unstable, meaning the instability grows on a dynamic timescale and does not depend on any kind of diffusion. Now suppose that there is diffusion of heat but not of salt. Consider what happens in each of the three dynamically stable scenarios when a fluid package is displaced from its equilibrium position. This analysis should reveal that one of the three remains stable; one should become 'overstable' (oscillations of growing amplitude) and the third should result in so-called *salt fingers*. Perform an experiment to examine one of these scenarios in a coffee cup—arrange for there to be more milk at the bottom (with a smooth milk gradient). It can help to gently heat from below. You should see the formation of distinct layers separated by discontinuities in the milk concentration. Exactly how a linearly overstable equilibrium should lead to such a 'staircase' is a field of active research.

5.6 **Length scales in Fourier space**

Using the programming language you find most convenient, construct a sawtooth function that repeats itself in the following way: a rising phase of width $a$, followed by a much narrower falling phase of width $b$ (or even better, with widths distributed in a Gaussian manner around $a$ and $b$). To avoid sharp transitions, make the rising and falling phases curved like a sine curve. Take the Fourier transform of this function. What length scales do you see in the function, and what length scales do you see in Fourier space?

5.7 **Gravitational collapse**

Building on the analysis in section 5.8, show that the Jeans length can also be estimated by equating the time taken for a sound wave to travel a certain distance to the freefall time over that distance. Furthermore, by considering a spherical cloud (of constant density, to simplify matters), show that the Jeans length is simply the size of the cloud in which the thermal and gravitational energies are comparable. Comment on the meaning of this, in terms of the virial theorem. Finally, calculate the Jeans length under typical conditions in the interstellar medium. For example, in much of the interstellar medium $c = 10$ km s$^{-1}$ and $\rho = 10^{-24}$ g cm$^{-3}$, and in dense clouds $c = 1$ km s$^{-1}$ and $\rho = 10^{-21}$ g cm$^{-3}$.

# References

Barber J L, Germann T C, Huang Z, Carles P, Kadau K, Rosenblatt C and Alder B J 2007 The importance of fluctuations in fluid mixing *Proc. Natl. Acad. Sci.* **104** 7741

Braithwaite J and Cavecchi Y 2012 A numerical magnetohydrodynamic scheme using the hydrostatic approximation *Mon. Not. R. Astron. Soc.* **427** 3265–79

Ji H, Burin M, Schartman E and Goodman J 2006 Hydrodynamic turbulence cannot transport angular momentum effectively in astrophysical disks *Nature* **444** 343–6

Kolmogorov A 1941 The local structure of turbulence in incompressible viscous fluid for very large Reynolds' numbers *Dokl. Akad. Nauk SSSR* **30** 301–5

Lumley J 1992 Some comments on turbulence *Phys. Fluids* **4** 203

Mouri H, Hori A and Kawashima Y 2003 Vortex tubes in velocity fields of laboratory isotropic turbulence: dependence on the Reynolds number *Phys. Rev.* E **67** 016305

Nordlund Å, Stein R F and Asplund M 2009 Solar surface convection *Living Rev. Solar Phys.* **6** 2

Vincent A and Meneguzzi M 1991 The spatial structure and statistical properties of homogeneous turbulence *J. Fluid Mech.* **225** 1–20

Woitke P 2006 2D models for dust-driven AGB star winds *Astron. Astrophys.* **452** 537–49

Zimmermann M 2009 Numerische simulation der Rayleigh-Bénard-konvektion *MA thesis* (Universität Bonn, Germany)

# Chapter 6

## Shocks

Generally speaking, a 'shock' is something that comes without you having expected it. In hydrodynamics, information normally travels at the sound speed; if something comes at you at more than that speed, you are not going to know about it until it is right on top of you.

In general, shocks are a phenomenon where a discontinuity in the density and pressure of a fluid appears when there is supersonic motion of some kind. This phenomenon is in some sense non-fluid in that the relevant length scale is of the order of the mean-free path of the particles, but we can still derive useful results without considering microscopic processes.

## 6.1 Viscous versus pressure gradient force

Let us first examine in what situation the viscous force (normally first-order in $\lambda/L$, the ratio of mean free path to characteristic length scale of the system under consideration) is comparable to the pressure gradient force. Looking at the momentum equation

$$\frac{d\mathbf{u}}{dt} = -\frac{1}{\rho}\nabla P + \nu\nabla^2\mathbf{u} \qquad (6.1)$$

we see that the ratio of pressure to viscous force is

$$\frac{F_P}{F_{\text{visc}}} \sim \frac{P}{\rho L}\left(\frac{\nu U}{L^2}\right)^{-1} \approx \frac{Lc^2}{U\nu} \approx \frac{c}{U} \cdot \frac{L}{\lambda}, \qquad (6.2)$$

where $U$ is the typical velocity. The gas viscosity relation $\nu \approx c\lambda$ has been used. Clearly, while $L \gg \lambda$ the viscous force is relatively unimportant; however, if the motion becomes supersonic ($U > c$) and a discontinuity appears ($L \approx \lambda$) then the viscous force becomes important. In this situation we should expect kinetic energy to be dissipated into heat.

doi:10.1088/978-1-6817-4597-8ch6
6-1

## 6.2 The jump conditions

To derive relations between the pressure, density and velocity on either side of a discontinuity it is easiest to go into the inertial frame in which the shock is at rest. In the following, quantities on either side of the discontinuity have the subscripts 0 and 1. The fluid has velocity component $u$ perpendicular to the discontinuity. By consideration of mass conservation—mass entering and leaving the discontinuity per unit time per unit area—we have

$$\rho_0 u_0 = \rho_1 u_1. \tag{6.3}$$

Next we can consider the rate of change of momentum contained within a volume spanning the discontinuity, which in the steady state (in the case of unsteady flow, we consider a short interval of time) must vanish. Contributions to increase momentum come from the momentum of the fluid entering the volume as well as the pressure $P_0$ acting on the volume; these must be balanced by momentum loss on the other side:

$$\rho_0 u_0^2 + P_0 = \rho_1 u_1^2 + P_1. \tag{6.4}$$

We now need to use conservation of energy. Again considering a volume spanning the discontinuity and equating the rate of change of its energy, made up of internal and kinetic energy flux inwards and outwards plus '$P\,dV$' work done, to zero, we have

$$\rho_0 u_0 \left( \varepsilon_0 + \frac{u_0^2}{2} + \frac{P_0}{\rho_0} \right) = \rho_1 u_1 \left( \varepsilon_1 + \frac{u_1^2}{2} + \frac{P_1}{\rho_1} \right), \tag{6.5}$$

where $\varepsilon$ is the internal energy per unit mass, a function of pressure and density. Of course, this is just an expression of Bernouilli's equation (3.3)[1]. Finally, we can consider velocity parallel to the discontinuity, where conservation of momentum gives us

$$\rho_0 u_0 \mathbf{v}_0 = \rho_1 u_1 \mathbf{v}_1, \tag{6.6}$$

where $\mathbf{v}$ is the component of velocity parallel to the discontinuity.

Now, there are two possible types of solution, assuming both densities and pressures are non-zero. First, we can have $u_0 = u_1 = 0$. In this case, we see from (6.3) and (6.6) that $\rho$ and $\mathbf{v}$ are unconstrained on both sides. However from (6.4) we have $P_0 = P_1$. This kind of discontinuity is called a *tangential discontinuity*, and in many contexts (particularly if the fluids on either side are of a different type or origin) it is called a *contact discontinuity*. Note that if the parallel velocities are not equal, i.e. if $\mathbf{v}_0 \neq \mathbf{v}_1$, then the flow is generally unstable (see section 5.3).

---

[1] Here we are simply equating the inward and outward energy fluxes on two surfaces of a volume. The energy flux through the sides of the volume is made to vanish by making the volume infinitesimally flat while still containing the discontinuity. For historical reasons this is called a *pillbox*.

The second solution has non-zero perpendicular velocities, and is called a *shock* or *shock wave*. From (6.3) and (6.6) we see that $\mathbf{v}_0 = \mathbf{v}_1$. Substituting (6.3) into (6.5) gives

$$h_0 + \frac{u_0^2}{2} = h_1 + \frac{u_1^2}{2}, \tag{6.7}$$

where internal energy $\varepsilon$ and $P/\rho$ have been brought together into enthalpy $h$. Now substituting (6.3) into (6.4) and rearranging we have

$$\rho_0 u_0 (u_0 - u_1) = P_1 - P_0. \tag{6.8}$$

To simplify the situation without loss of generality we now define the axes in such a way that both $u_0$ and $u_1$ are positive, so that they represent upstream and downstream velocities respectively. This leaves the following two possibilities: either $u_0 > u_1$, $P_0 < P_1$ and $\rho_0 < \rho_1$ or $u_0 < u_1$, $P_0 > P_1$ and $\rho_0 > \rho_1$. It is left as an exercise for the student to show formally that these two possibilities represent an increase and a decrease in entropy. Since we know that entropy must increase, only one set of solutions is permissible—namely that $u_0 > u_1$. This means that the gas is compressed and heated on its passage through the shock. The energy to do this can only come from the decrease in kinetic energy. In fact, the opposite of this would apparently violate the second law of thermodynamics, according to which it is impossible to construct a system whose sole result is the conversion of energy from heat to kinetic energy.

The change of state of the gas happens via microscopic processes in the shock, which generally has a thickness of order the mean free path of the particles. An interesting feature is that the change in the thermodynamic state of the gas is determined entirely by the macroscopic quantities on either side; the shock itself can be thought of as adjusting itself to meet the external requirements placed upon it, regardless of the fluid's microscopic properties. For instance, a shock can be passed at a given speed through two fluids which are identical except for their viscosities; in the less viscous fluid the shock discontinuity will become thinner to allow it to dissipate the same energy as in the more viscous fluid, and it is impossible to tell the difference between the two if one just measures the macroscopic quantities. This behaviour is also seen for instance in magnetic reconnection.

To look at the interdependence of the variables in (6.3), (6.4) and (6.7) it is first helpful to express the enthalpy in terms of pressure and density. To make matters simpler, we look from now on at an ideal gas with $h = (P/\rho)\gamma/(\gamma - 1)$. After some algebra, we arrive at:

$$\frac{P_1}{P_0} = 1 + \frac{2\gamma}{\gamma + 1}\left(M_0^2 - 1\right), \tag{6.9}$$

where $M \equiv u/c$ is the Mach number. Note that the Mach numbers on either side of the shock are the velocities as fractions of the sound speed on the respective side. The fluid enters the shock supersonically (remember that we defined the directions such that $P_1 > P_0$). After more algebra we have

$$M_1^2 = \frac{(\gamma - 1)M_0^2 + 2}{2\gamma M_0^2 + 1 - \gamma}. \tag{6.10}$$

It is easily verified that $M_0 > 1$ and $M_1 < 1$ (except in the trivial solution where all quantities are the same on both sides). Therefore, the material enters supersonically and exits subsonically. As $M_0$ tends towards unity from above, $M_1$ tends towards unity from below. Furthermore, as $M_0 \to \infty$, $M_1 \to (\gamma - 1)/2\gamma$, so there is a limit to the conversion of kinetic energy into heat. The density and velocity ratios are

$$\frac{\rho_1}{\rho_0} = \frac{(\gamma + 1)M_0^2}{(\gamma - 1)M_0^2 + 2} = \frac{u_0}{u_1}. \tag{6.11}$$

An important point to note at this juncture is the limit on the compression factor of $(\gamma + 1)/(\gamma - 1)$, which is equal to 4 for a monatomic gas, in which case no more than 15/16 of the kinetic energy can be converted into heat.

## 6.3 Contexts

Where do we see shocks? A good natural example would be the shock we hear as thunder as the air in a lightening bolt reacts to a sudden and large change in pressure. Supersonic aeroplanes also produce shocks.

Often when things catch fire and explode there is a shock. It is useful here to examine briefly the distinction between *deflagration* and *detonation*. In a deflagration, the burning spreads by transfer of heat. A fluid element ignites and the resulting heat is transferred via thermal diffusion to the next fluid element, bringing it up to ignition temperature. The burning spreads subsonically. In a detonation on the other hand, the burning spreads supersonically; the fluid is brought up to ignition temperature by a shock. Deflagrations can happen rather quickly; examples include the gunpowder in guns and fireworks as well as petrol in an internal combustion engine. Detonation is what occurs for instance in most bombs. One thing that is not yet known about the explosion of white dwarfs (which is thought to result in type Ia supernovae) is whether it is deflagration or detonation.

As we saw in section 3.4, in *Bondi accretion* a star accretes material (spherically-symmetrically) and the material becomes supersonic at some radius. What happens next? Whether the star is a young star accreting from a large gas cloud or a neutron star accreting from a companion, the material hits an *accretion shock*, a standing shock somewhat above the star's surface. It generally heats up sufficiently to emit x-rays or $\gamma$-rays.

Another example of a relatively stationary shock is where the solar wind meets the boundary of the Earth's ionosphere. Another classic example of a shock in astrophysics is that created by the wind of a star, leading to a particular structure. The wind flies supersonically outwards from the star, and passes through a *reverse shock*, whereupon it continues to expand at a much slower rate. Beyond there is the *contact discontinuity*, separating shocked wind material from interstellar medium (ISM). Generally this ISM has also been shocked, upon passing through a *forward shock* which flies outwards supersonically into the surrounding ISM. We therefore have three concentric spherical discontinuities. In reality of course the situation is normally complicated by the fact that the star is moving relative to the ISM and so the surfaces are not spheres. Note also that the situation can be complicated by

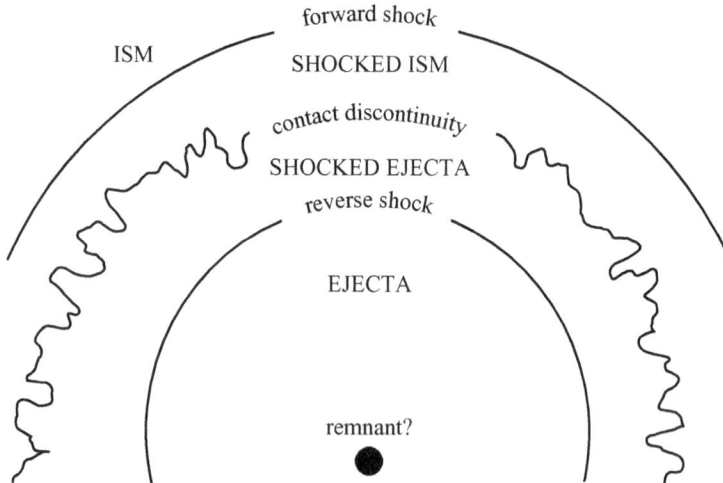

**Figure 6.1.** The structure of the shells around a recently-exploded star. Note that the contact discontinuity is Rayleigh–Taylor unstable.

radiation from the central object, which flies outwards ahead of the wind and ionises the surrounding material, also creating a shock.

Supernovae produce shock waves similarly to stellar winds, with reverse shock, contact discontinuity and forward shock (state figure 6.1). In this context it is possible to model the structure as a function of radius and time in terms of a similarity variable, a combination of radius and time, in such a way that the variables at any radius and time can be given as a function of just this similarity variable. Such a solution in supernova and also the detonation of nuclear bombs is often called the Sedov blast wave.

Note that in some astrophysical contexts the maximum density ratio across the shock, equal to 4 in a monatomic gas, apparently doesn't apply: much higher density ratios are observed. The trick here is that the gas cools (by radiation) after it passes through the shock; from a distance you can't distinguish the shock itself from the cooling layer[2].

The above examples are all related to sound waves; indeed in the analysis above we considered a system which could host sound waves and no other kind of wave. There are however many other types of shock. Note for instance the similarity between the shock wave arising from a supersonic aircraft and the wake made by a boat or duck. The related wave here is the deep water wave. These being dispersive waves, the properties are somewhat different from sound-related shocks. We look in the next section at a shock where the related wave is the shallow water wave.

## 6.4 Hydraulic jumps

It was shown in section 5.1 that shallow water waves (i.e. where $\lambda \gg h$) propagate with a speed $\sqrt{gh}$ where $h$ is the depth of the water, and that waves with shorter

---

[2] This phenomenon has given rise to some confusing jargon: an 'adiabatic shock' is a shock that lacks such cooling.

wavelength propagate more slowly in water of the same depth. This is therefore the maximum speed at which disturbances can propagate. Now imagine a discontinuity in the depth of water parallel to the line $x = 0$, with depth $h_0$ on the left and $h_1$ on the right. Relative to the discontinuity, the water is flowing into the discontinuity from the left at speed $u_0$ and away from it on the right at speed $u_1$. Mass conservation gives

$$h_0 u_0 = h_1 u_1 \tag{6.12}$$

while momentum conservation gives

$$h_0 u_0^2 + \frac{1}{2} g h_0^2 = h_1 u_1^2 + \frac{1}{2} g h_1^2, \tag{6.13}$$

where the first term on each side represents momentum advected into and out of a volume containing the shock and the second term on each side represents the pressure exerted on that volume. (The density of the water obviously drops out of both of these equations.) If we know $h_0$ and $u_0$ then we can calculate $h_1$ and $u_1$. If we now considered the flow of kinetic and potential energy ($hu^2/2 + gh^2/2)\rho u$ into and out of the volume (per unit length discontinuity), we would find the problem had become overconstrained. This is because our water has only one equivalent of a thermodynamic variable, $h$, compared to two degrees of freedom in the gas considered in the previous section. But does this mean energy is not conserved? The answer is that energy is converted into a form we have not considered, i.e. disordered kinetic and heat. The rate at which the energy is converted is

$$\begin{aligned} q &= \frac{\rho u_0}{2}\left(h_0 u_0^2 + g h_0^2\right) - \frac{\rho u_1}{2}\left(h_1 u_1^2 + g h_1^2\right) \\ &= \frac{\rho g h_0 u_0 \left(h_0^2 + h_1^2\right)(h_1 - h_0)}{4 h_0 h_1}, \end{aligned} \tag{6.14}$$

where (6.12) and (6.13) have been used. Since this energy must be positive, we can see that $h_1 > h_0$. With the help of some algebra it is also possible to show that $u_0 > \sqrt{g h_0}$ and $u_1 < \sqrt{g h_1}$. The water enters the jump at a speed faster than the wave propagation speed and exits below this speed.

## Exercises

### 6.1 Compressibility and energy
We saw above that there is a limit to the ratio of densities on either side of the shock, equal to 4 in a monatomic gas. Show that the temperature ratio is

$$\frac{T_1}{T_0} = 1 + \frac{2(\gamma - 1)}{(\gamma + 1)^2} \cdot \frac{\left(\gamma M_0^2 + 1\right)\left(M_0^2 - 1\right)}{M_0^2}, \tag{6.15}$$

and find a similar expression for the pressure ratio. How do these ratios behave as the Mach numbers tend to unity and infinity?

**6.2 The simplest kitchen sink experiment**

Find a sink and turn on the tap. Do you see a discontinuity?

**6.3 Steepening of sound waves**

During the analysis of sound waves in section 2.2, we neglected all second-order terms. Without actually doing the analysis again keeping these terms, consider their effect qualitatively. What might be the end result? (Hint: consider the difference in the sound speed between regions of compression and rarefaction.) Should the same happen with shallow water waves?

# Chapter 7

## Vorticity and rotating fluids

In this chapter we look at vorticity in more detail as well as various phenomena in rotating systems. First, recall that in section 2.4.2 vorticity was defined as $\omega \equiv \nabla \times \mathbf{u}$ and that circulation $\Gamma$ was defined as the integral of velocity around a (co-moving) closed loop. Recall also that circulation is conserved in an inviscid, barotropic flow where body forces are conservative.

### 7.1 Vortices

Here, we take a look at vorticity generation, vortex tubes and the interaction of vortices. Imagine two basic velocity fields, with

$$u_\theta = \frac{1}{2}\omega\varpi \qquad \text{and} \qquad u_\theta = \frac{\Gamma}{2\pi\varpi}, \tag{7.1}$$

where the former represents solid-body rotation and the latter is irrotational (i.e. zero vorticity) everywhere except for a singularity on the axis. ($\varpi$ is the cylindrical radius.) An everyday example of solid-body rotation would be a cup of coffee which has come into a rotational equilibrium inside a microwave. The latter kind, called a *line vortex*, does not really exist in nature; rather, the singularity is replaced with an inner cylinder of solid-body rotation, surrounded as before by irrotational flow. This is called a *Rankine vortex*; it can alternatively be described as a cylinder of constant vorticity (a *vortex tube*) surrounded by zero vorticity. A tornado can well be approximated by this kind of vortex (see figure 7.1); note that the radius of the inner region is very small compared to the sky and cloud from which the tornado forms, so if one is only interested in the longer length scales one can make the approximation of a perfect line vortex.

We can draw vorticity lines in the same way as we can draw streamlines. Note that the divergence of vorticity is zero. Like magnetic fields in a conducting fluid, vortex lines can be thought of as being 'frozen' into the fluid. Imagine a patch $S$ on the surface of a vortex tube. Since the vorticity is everywhere parallel to this surface,

doi:10.1088/978-1-6817-4597-8ch7

**Figure 7.1.** A tornado.

the circulation around its perimeter is zero. As the fluid moves around, Kelvin's circulation theorem tells us that the circulation must always remain zero; therefore so must the vorticity perpendicular to the surface, so that the comoving patch must remain on the surface of the vortex tube. We conclude that vortex lines are frozen into the fluid. This is analogous to the freezing of magnetic field lines into a conducting fluid (section 8.5).

We now consider very briefly the interaction of vortices with each other and with solid boundaries. First we approximate a vortex with a line vortex as described above. Now, since the flow around a line vortex is irrotational, we can write $\mathbf{u} = \nabla \phi$ where $\phi$ is a scalar potential. When two or more vortices are present, we can simply add together the scalars belonging to each vortex in isolation to produce the resultant velocity field. Recalling from above that vortex lines are 'frozen' into the fluid, we see that each individual vortex must move according to the sum of the velocity fields belonging to each of the other vortices in isolation.

First let us consider two parallel vortices of equal magnitude and sense. Associated with each one is a velocity field in which the other moves around; in this case they encircle each other. Far away from the vortices, the two velocity fields cancel each other out. If, on the other hand, the two vortices have a circulation of the opposite directions, the pair will move together in a straight line perpendicular to the line between them. Anyone who has tried rowing will be familiar with vortex pairs created by an oar travelling surprisingly large distances behind the boat.

Now imagine a vortex ring such as that created by a skillful smoker. Each section of the ring moves according to the velocity field associated with the rest of the ring, with the result that the ring propagates forwards much further than one would otherwise expect. Another example of a vortex ring is shown in figure 7.3.

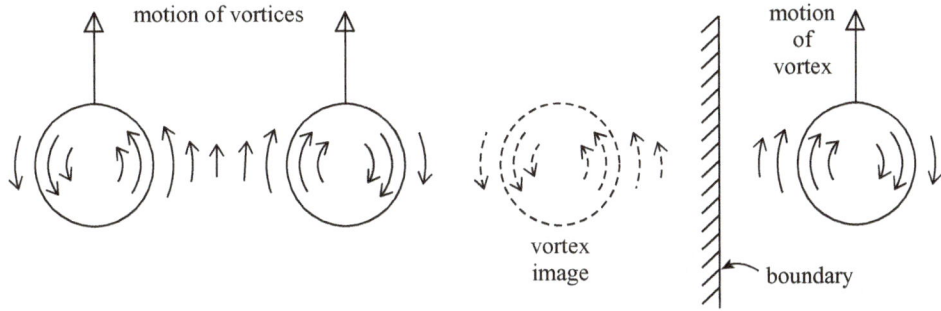

**Figure 7.2.** Left: a pair of vortices rotating in opposite senses. Each vortex moves in the velocity field of the other, with the result that they move in the same direction. Right: a single vortex near a boundary. From symmetry we see that the velocity field must be the same as in the case of the pair of vortices, so the vortex moves along the boundary. (This assumes a free-slip boundary, which is reasonable if the Reynolds number is high so that the boundary layer is thin.)

**Figure 7.3.** Fascination with vortices is not restricted to homo sapiens (https://www.pinterest.de/pin/204702745534974759/).

As a vortex approaches a boundary, we can predict what will happen with the method of images where one imagines removing the boundary and placing an image vortex on the other side (see figure 7.2). In the case of a single vortex near a boundary, we can recreate the flow with an image vortex of opposite spin, since the velocity field where the boundary once was is now constrained to be parallel to it. Of course in reality there will be a small difference between the two flows, namely that there will be a thin boundary layer, but that can ignored in the bulk of the volume. So, a single vortex near a wall will move parallel to the wall.

The behaviour of vortices is endlessly fascinating; the reader is heartily advised to look them up on his or her favourite video-streaming website.

## 7.2 The vorticity equation

Let us take the curl of the momentum equation (4.8), taking care first to write out the Lagrangian time derivative as Eulerian time derivative and advective term:

$$\frac{\partial \omega}{\partial t} + \nabla \times (\mathbf{u} \cdot \nabla \mathbf{u}) = - \nabla \times \left( \frac{1}{\rho} \nabla P \right) + \nabla \times \mathbf{g} + \nabla \times (\nu \nabla^2 \mathbf{u}),$$

$$\frac{\partial \omega}{\partial t} + \nabla \times (\omega \times \mathbf{u}) = - \nabla \left( \frac{1}{\rho} \right) \times \nabla P + \nu \nabla^2 \omega, \tag{7.2}$$

$$\frac{d\omega}{dt} = \frac{1}{\rho^2} \nabla \rho \times \nabla P - \omega (\nabla \cdot \mathbf{u}) + (\omega \cdot \nabla) \mathbf{u} + \nu \nabla^2 \omega,$$

where we have used the fact that the gravitational force is conservative (and assumed that any other body forces present are also conservative) as well as some vector calculus identities (see appendix A.3). This equation is called the *vorticity equation*. The first term on the right vanishes in a barotropic flow, the next two terms on the right contain vorticity, and the last term is the viscosity, so that we recover the conclusion (see sections 2.4.1 and 2.4.2) that if vorticity is zero everywhere at some point in time, it is zero at all other times, provided that the flow is barotropic and inviscid and that the body forces are conservative.

It is informative to examine the physical meaning of each of the terms on the right-hand side of (7.2). On the right-hand side we have:

(a) The baroclinic term. The pressure acts normal to the surface of a fluid element; imagine that the fluid element is spherical: the total pressure force, equal to the sum of the pressure force on surface elements, acts through the centre of the fluid element and is parallel to the gradient of pressure. The fluid element can be made to rotate if the centre of mass does not lie on the line of net pressure gradient force; however if $\rho = \rho(P)$ then this is impossible. Therefore the pressure gradient force can only bring about a linear acceleration of the fluid element and cannot act as a torque there-upon: hence the constancy of circulation. Note that vorticity is an expression of the rotation of a fluid element about its own centre of mass, not about any other point—fluid elements with zero vorticity are still free to move in circles around each other. This is the reason that gravity cannot make the fluid element rotate, as it acts through the centre of mass.

A good example of where the baroclinic term can generate torque is after heating in the atmosphere, for instance after an explosion has created a hot bubble. If the bubble has uniform temperature then vorticity is created at its boundary when the bubble rises, and if the bubble has a temperature profile which gradually decreases outwards then vorticity is generated throughout the volume; the result of this is that the bubble deforms into a rising vortex ring.

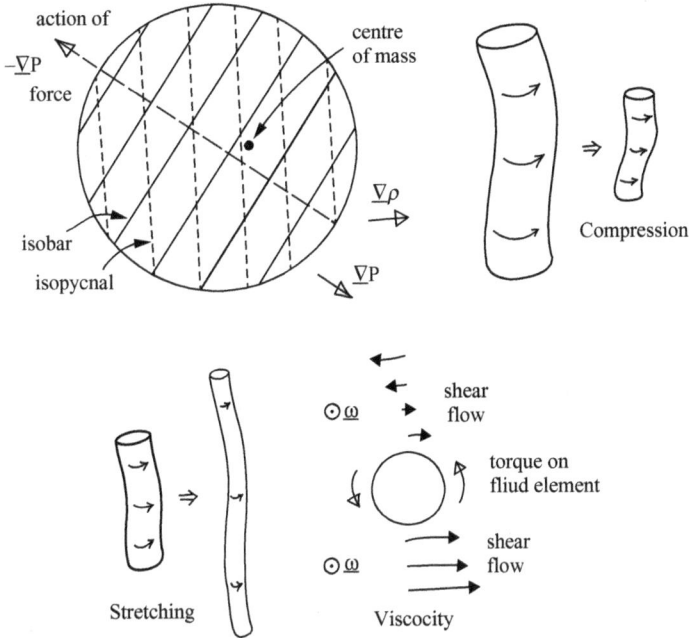

**Figure 7.4.** Four different ways to generate vorticity. From left to right: the baroclinic term, where if $\nabla P$ and $\nabla \rho$ are not parallel the pressure gradient force on a fluid element does not go through the centre of mass; compression of a vortex reduces the moment of inertia and so to conserve angular momentum, the angular velocity must increase; likewise stretching at constant volume also results in spin-up; viscosity can make a fluid element rotate if there is vorticity present in the neighbouring fluid. Image Credit: NASA/GSFC/JPL, MISR Team.

(b) The second term simply represents conservation of angular momentum as the fluid is compressed or expands; clearly the vorticity should increase if the density increases, which happens if the divergence of the velocity is negative, hence the minus sign above. One can also think in terms of the vortex lines being squashed together when the density rises, increasing the density of vortex lines and hence the vorticity.

(c) The third term represents conservation of angular momentum during 'stretching' of vortices. If a fluid element is stretched out along the direction of the vorticity while its volume remains constant, then the moment of inertia of the fluid element has decreased and so the angular velocity and vorticity must increase. Again one can think in terms of vortex lines being squashed closer together and the vorticity, which is the density of vortex lines, increasing. The stretching of vortex tubes seems to be fundamental to the phenomenon of turbulence (section 5.7). Note an important difference though between two- and three-dimensional flow—it is impossible to stretch a vortex in two dimensions.

(d) The viscous term, which has the same form as that in the original momentum equation, simply causes vorticity to diffuse from local maxima to minima. It causes an isolated vortex tube of the kind described in the previous section to become broader. It also generates vorticity in boundary layers.

These mechanisms are illustrated in figure 7.4. In addition, non-conservative body forces can also generate vorticity; an example is the Coriolis force, which we examine in section 7.5.

## 7.3 The momentum equation in a rotating frame of reference

In a frame of reference which is rotating with angular velocity $\mathbf{\Omega}$ with respect to the non-rotating, inertial frame, the Lagrangian time derivatives of scalar quantities such as density and temperature must of course be the same. However, velocity is frame dependent and so generally the Lagrangian derivative of velocity is different in the two frames. For instance, a fluid element which is stationary in the rotating frame must be experiencing an acceleration in the inertial frame. It turns out (after a lengthy derivation) that the comoving derivatives of velocity in the two frames are related by

$$\left(\frac{d\mathbf{u}_I}{dt}\right)_R = \left(\frac{d\mathbf{u}_R}{dt}\right)_I - \mathbf{\Omega} \times \mathbf{\Omega} \times \mathbf{r}_R - 2\mathbf{\Omega} \times \mathbf{u}_R - \frac{d\mathbf{\Omega}}{dt} \times \mathbf{r}_R, \quad (7.3)$$

where the subscripts I and R denote inertial and rotating frames. The difference between the two comoving derivatives is the three acceleration terms on the right-hand side, the origin of which can be understood intuitively in the following way:

(a) The term $-\mathbf{\Omega} \times \mathbf{\Omega} \times \mathbf{r}_R$ is the centrifugal acceleration: to make anything move in a circle (i.e. accelerate towards the centre of the circle) it is necessary to provide an inwards-directed force (for instance from gravity) of magnitude $\varpi\Omega^2$ per unit mass where $\varpi$ is the distance from the axis of rotation. In the rotating frame in which the fluid element is stationary, there is no longer any acceleration towards the centre. The original inwards force therefore must be balanced by an extra force directed outwards, the centrifugal force. Note that the centrifugal force can be written $\hat{\boldsymbol{\varpi}}\varpi\Omega^2$ where $\hat{\boldsymbol{\varpi}}$ is the cylindrical radius unit vector.

(b) Unlike the centrifugal force, the Coriolis acceleration $-2\mathbf{\Omega} \times \mathbf{u}_R$ arises only when fluid is *moving* within the rotating frame; it can be thought of as accounting for conservation of angular momentum. If a fluid element moves in the azimuthal direction, i.e. with a different angular velocity to the frame of reference, it experiences a different centrifugal acceleration from the frame and so accelerates in the (cylindrical) radial direction relative to the frame. Similarly, if a fluid element moves to a larger radius while preserving its angular momentum it must attain a smaller angular velocity and therefore begin to drift backwards in the azimuthal direction relative to the rotating frame. One can think of standing on the north pole and firing a projectile horizontally—in the inertial frame the object moves in a straight line but since the Earth is rotating, an observer on the ground will see the object curve towards the west. Note that there is no component of the Coriolis acceleration in the direction of the angular velocity $\mathbf{\Omega}$.

(c) The third term represents the apparent acceleration a fluid element experiences when the underlying rotation of the reference frame changes.

In most situations we use a reference frame with constant $\mathbf{\Omega}$ and can drop this term, as we do in all of the following.

Putting this together into a rotating-frame momentum equation, we have

$$\frac{d\mathbf{u}}{dt} = -\frac{1}{\rho}\nabla P - \nabla\Phi_{\text{eff}} - 2\mathbf{\Omega} \times \mathbf{u} + \nu\nabla^2\mathbf{u}. \tag{7.4}$$

The viscous term has been simplified in the usual way and $\mathbf{r}$ and $\mathbf{u}$ are the position and velocity vectors in the rotating frame, the subscript R having been dropped. The centrifugal term has been absorbed into the gravitational potential, which is possible because it is conservative:

$$\mathbf{g}_{\text{eff}} \equiv \mathbf{g} - \mathbf{\Omega} \times \mathbf{\Omega} \times \mathbf{r} = -\nabla\Phi_{\text{grav}} + \nabla\left(\frac{1}{2}\Omega^2\varpi^2\right)$$

$$= -\nabla\Phi_{\text{eff}} \quad \text{where} \quad \Phi_{\text{eff}} \equiv \Phi_{\text{grav}} - \frac{1}{2}\varpi^2\Omega^2. \tag{7.5}$$

## 7.4 The centrifugal force and the von Zeipel paradox

In rotating systems the Coriolis force is usually more interesting than the centrifugal force, and indeed the rest of this chapter is dedicated to its effects. Beforehand though, we look at an effect of the centrifugal force. Now, since it can be absorbed into the gravitational potential the centrifugal force merely produces a change in the shape of equipotential surfaces. This is the reason that rotating stars and planets are flattened (see also exercise 7.9). Rotational flattening is clearly visible with a small telescope trained on Jupiter or Saturn (see figure 7.5).

**Figure 7.5.** The rotational flattening of Jupiter (http://www.azastronomy.com/solar-system/jupiter-2003.html, copyright Richard Jacobs) and Saturn (http://www.ozscopes.com.au/what-can-you-expect-to-see-with-a-tele scope.html) is clearly visible in these excellent amateur images. Image Credit: NASA/GSFC/JPL, MISR Team.

An interesting effect of the centrifugal force in stars was discovered by von Zeipel (1924). Now, in a static equilibrium in a rotating frame of reference we must have

$$\frac{1}{\rho}\nabla P = -\nabla\Phi_{\text{eff}} \tag{7.6}$$

from which we see, by taking the curl, that

$$\nabla\rho \times \nabla P = \mathbf{0}. \tag{7.7}$$

So we must have a barotropic relation $\rho = \rho(P)$, and so the contours of pressure, density, temperature and effective potential $\Phi_{\text{eff}}$ must all coincide. However, the flux of heat is proportional to $-\nabla T$, so the flux must be greater at the poles of the star than around the equator, and the divergence of this flux must be equal to the rate of nuclear energy generation, which is a function of the local thermodynamic state but not of its gradient. Thus the rotation places demands on the nuclear energy generation which in general cannot be met; there is therefore no static equilibrium in a rotating star. This is called the *von Zeipel paradox*.

The solution might be complicated. A popular solution is *Eddington–Sweet circulation* (see Sweet 1950) where a meridional flow exists to move material outwards and inwards at the poles and equator, respectively. Busse (1982) then pointed out that this circulation is a solution of the heat equation but not of the momentum equation. Zahn (1992) suggested instead a solution in terms of baroclinic instability, a convection-like instability present in a gravitationally-stratified fluid where normal convection is absent but where the gradients of pressure and density are not parallel (see section 5.6). This creates flows which result in mixing on spherical surfaces, but with little exchange between different radii. Consequently all quantities should be roughly constant on spherical surfaces and depend only on the radial coordinate. This result was inevitably greeted with enthusiasm by the stellar evolution community, who had always assumed spherical symmetry anyway, for practical reasons. There is much evidence however that the matter is not yet satisfactorily resolved. While many published stellar evolution calculations claim to include rotational mixing, one should be aware that this involves using a mixing theory which is almost certainly not the full story plus a free parameter whose value is chosen empirically and differently for different stars.[1]

## 7.5 The vorticity equation in a rotating frame

In section 7.2 we took the curl of the momentum equation and derived equations for the Eulerian and Lagrangian time derivatives of vorticity. We now extend this procedure to a rotating frame of reference. We include therefore the Coriolis force in

---

[1] There is a parallel here with convection and mixing length theory. The assumptions on which mixing length theory is based are known to be wrong (there is not much mixing, the assumed length scale doesn't exist, etc), but in the absence of another convenient prescription it is included in stellar evolution calculations to describe convective zones. Not surprisingly the results of these calculations are sometimes frighteningly sensitive to the chosen value of the 'mixing length parameter'.

the momentum equation, but not the centrifugal force, which can usually safely be integrated into gravity as in (7.5). We obtain

$$\frac{\partial \omega}{\partial t} + \nabla \times (\mathbf{u} \cdot \nabla \mathbf{u}) = -\nabla \times \left(\frac{1}{\rho}\nabla P\right) + \nabla \times \mathbf{g}$$
$$-\nabla \times 2\mathbf{\Omega} \times \mathbf{u} + \nabla \times (\nu\nabla^2\mathbf{u}),$$

$$\frac{\partial}{\partial t}(\omega + 2\mathbf{\Omega}) + \nabla \times ((\omega + 2\mathbf{\Omega}) \times \mathbf{u}) = -\nabla\left(\frac{1}{\rho}\right) \times \nabla P + \nu\nabla^2(\omega + 2\mathbf{\Omega}), \quad (7.8)$$

$$\frac{\mathrm{d}}{\mathrm{d}t}(\omega + 2\mathbf{\Omega}) = -\nabla\left(\frac{1}{\rho}\right) \times \nabla P - (\omega + 2\mathbf{\Omega})\nabla \cdot \mathbf{u}$$
$$+ [(\omega + 2\mathbf{\Omega}) \cdot \nabla]\mathbf{u} + \nu\nabla^2(\omega + 2\mathbf{\Omega}),$$

where on the second and third lines $\mathbf{\Omega}$ could be taken inside the time and spatial derivatives because they are zero. Comparing with (7.2) we see that vorticity has simply been replaced by $\omega + 2\mathbf{\Omega}$, the *absolute vorticity*. In rotating systems $\omega$ is referred to as the *relative vorticity*.

There is a logic to the equivalence of $\omega$ in a non-rotating frame and $\omega + 2\mathbf{\Omega}$ in a rotating frame if we consider that $2\mathbf{\Omega}$ is the 'vorticity of the rotating frame', recalling from section 2.4.2 that vorticity is double the angular velocity. The equivalence is also valid for Kelvin's circulation theorem (section 2.4.1). Defining:

$$\Gamma \equiv \oint (\mathbf{u} + \mathbf{\Omega} \times \mathbf{r}) \cdot \delta\mathbf{s}, \quad (7.9)$$

we find that $\mathrm{d}\Gamma/\mathrm{d}t = 0$ in the absence of baroclinicity, viscosity and non-conservative body forces, just as before. Applying Stokes' theorem we have

$$\frac{\mathrm{d}}{\mathrm{d}t}\int(\omega + 2\mathbf{\Omega}) \cdot \mathrm{d}\mathbf{S} = 0. \quad (7.10)$$

We see from this a mechanism for generating (relative) vorticity where there was none in the non-rotating frame; if a fluid element changes its extent in the plane perpendicular to $\mathbf{\Omega}$ it will start to spin relative to the rotating frame *even if it did not do so initially*. An example of this is when heating on the Earth causes the air over a region to warm up and expand laterally, such as happens in a hurricane.

## 7.6 Inertial waves

We have seen previously how the pressure gradient force and gravity can both provide a restoring force for waves; in this section we look at waves with the Coriolis force as the restoring agent. To simplify the equations (avoiding solving also for sound and gravity waves at the same time as the inertial waves we are investigating) we make the assumption of constant density, meaning that we can automatically ignore gravity since gravity has no interesting effect on a constant-density fluid except at the surface; we look here at waves internal to the fluid. The momentum equation is

$$\frac{\partial \mathbf{u}}{\partial t} + (\mathbf{u} \cdot \nabla)\mathbf{u} = -\frac{1}{\rho}\nabla P - 2\mathbf{\Omega} \times \mathbf{u} + \hat{\varpi}\varpi\Omega^2. \tag{7.11}$$

We now make the usual assumption when looking at waves of small amplitude, so that we can drop the second term on the left. Taking the curl of the remaining terms we have

$$\frac{\partial}{\partial t}\nabla \times \mathbf{u} = -2\nabla \times (\mathbf{\Omega} \times \mathbf{u}) \tag{7.12}$$

since the curls of the centrifugal and pressure gradients forces are zero. In fact, in many situations the centrifugal force can be ignored since it can be expressed as the gradient of a scalar; as we saw above, in situations with gravity it can simply be added to the gravitational potential which often can be removed completely by an adjustment to the definition of the vertical axis, in a local analysis. Now, the term on the right of (7.12) can be rewritten in a more convenient form; with the help of a vector calculus identity $\nabla \times (\mathbf{\Omega} \times \mathbf{u}) = \mathbf{\Omega}(\nabla \cdot \mathbf{u}) - (\mathbf{\Omega} \cdot \nabla)\mathbf{u}$, and noting that the incompressible continuity equation means that the first of these two terms vanishes. Finally we set the rotation axis along the $z$-axis and write

$$\frac{\partial}{\partial t}\nabla \times \mathbf{u} = 2\Omega\frac{\partial \mathbf{u}}{\partial z}. \tag{7.13}$$

As before, we now assume a solution of the form $\exp[i(\mathbf{k} \cdot \mathbf{r} - \omega t)]$; the continuity equation gives

$$\mathbf{k} \cdot \mathbf{u} = 0 \tag{7.14}$$

and the momentum equation (7.13) becomes

$$\omega\mathbf{k} \times \mathbf{u} = 2\Omega ik_z\mathbf{u}. \tag{7.15}$$

We now take the curl of both sides and noting that $\mathbf{k}$ and $\mathbf{u}$ are perpendicular we can write $\mathbf{k} \times \mathbf{k} \times \mathbf{u} = -k^2\mathbf{u}$:

$$-\omega k^2\mathbf{u} = 2\Omega ik_z\mathbf{k} \times \mathbf{u}, \tag{7.16}$$

which we can compare with the previous form to give

$$\omega = 2\Omega\frac{k_z}{k} = 2\Omega\cos\theta, \tag{7.17}$$

where $k$ is the magnitude of the wavevector $\mathbf{k}$ and $\theta$ is the angle between the wavevector and the rotation axis $\mathbf{\Omega}$. This is a similar dispersion relation to that of internal gravity waves (5.49) in that the frequency of the oscillations depends only on the direction of the wavevector and not on its magnitude. Here, the frequency of the oscillations goes to zero when the wavevector is perpendicular to the rotation axis, i.e. when the velocity field is parallel to it. It is of course obvious from the form of the Coriolis force that motion parallel to the rotation axis experiences no restoring force.

Another interesting feature of these waves is that the energy is entirely kinetic, rather than being converted back and forth between two different forms. Instead one has to think here of energy conversion back and forth between kinetic energy of motion in two different directions.

## 7.7 The Taylor–Proudman theorem

Imagine motions in a rotating fluid with characteristic length-scale, time-scale and velocity $L$, $T$ and $U$. Looking at the sizes of the various terms in the momentum equation we have

$$\frac{\partial \mathbf{u}}{\partial t} + (\mathbf{u} \cdot \nabla)\mathbf{u} = -\frac{1}{\rho}\nabla P - 2\mathbf{\Omega} \times \mathbf{u} \tag{7.18}$$

$$\frac{U}{T} \qquad \frac{U^2}{L} \qquad \frac{\delta P}{\rho} \qquad \Omega U \tag{7.19}$$

where the centrifugal force has been ignored as it can be absorbed into the equilibrium pressure gradient force. Furthermore we assume a relation $T \sim L/U$, typical of flows rather than waves, meaning that the two terms on the left-hand side are of comparable size. The ratio of the Coriolis to inertial terms is $\Omega T$; the inverse of this number is called the *Rossby number* after the Swedish physicist. At low Rossby number, i.e. $T^{-1} \ll \Omega$, the momentum equation can be approximated to

$$\frac{1}{\rho}\nabla P = -2\mathbf{\Omega} \times \mathbf{u}. \tag{7.20}$$

We see from this that the gradient of $P$ in the direction of the rotation axis vanishes; furthermore we see that this equation, which relates the velocity perpendicular to the rotation axis to the pressure gradient perpendicular to it, demonstrates that the gradient of the velocity field along the rotation axis also vanishes. Finally, taking the curl of this equation, we lose the left-hand side if the fluid is incompressible, and using a vector identity on the remaining term (again assuming incompressibility and therefore $\nabla \cdot \mathbf{u} = 0$) gives $(\mathbf{\Omega} \cdot \nabla)\mathbf{u} = 0$. In summary, if the rotation axis is parallel to the $z$ axis, we have

$$\frac{\partial w}{\partial z} = 0 \quad \text{and} \quad \frac{\partial u}{\partial x} + \frac{\partial v}{\partial y} = 0 \quad \text{and in general} \quad \frac{\partial}{\partial z} = 0. \tag{7.21}$$

This is called the *Taylor–Proudman theorem*, which is illustrated in figure 7.6. In the other extreme, in the limit of high Rossby number we may ignore the Coriolis force altogether.

Taylor–Proudman 'columns' are thought to exist in rotating astrophysical bodies which are convective. Instead of moving up and down in the radial direction, convective cells move up and down parallel to the rotation axis.

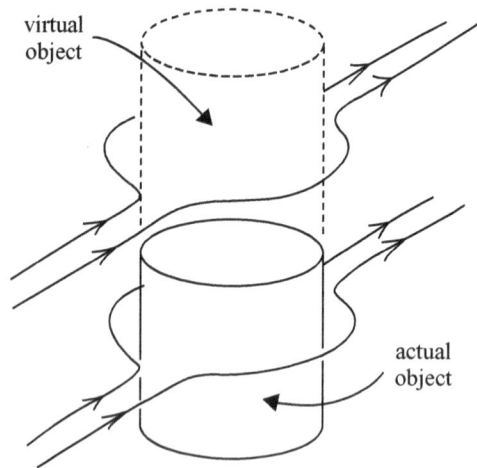

**Figure 7.6.** The flow of fluid around and above an obstacle in a rotating frame with low Rossby number.

## 7.8 The geostrophic approximation

As we have done before, by comparing sizes of various terms in the momentum equation we can, according to the context, simplify it by dropping all but the largest terms. In the following we look at the example of the Earth's atmosphere and oceans, but the same principles are applicable in many other astrophysical objects.

The Earth rotates considerably more slowly than the break-up spin (~2 h) and we can ignore the centrifugal force. Furthermore, the Earth's atmosphere and oceans are very thin compared to its horizontal extent, and the gravitational force is much stronger than inertia or the Coriolis force. Thus, for most purposes we can assume hydrostatic equilibrium and drop the vertical component of velocity from the equations, which amongst other things allows us to simplify the form of the Coriolis force. If the colatitude (defined as 0° at the north pole, 90° at the equator and 180° at the south pole) is $\theta$ then we define $f \equiv 2\Omega \cos \theta$ and the two horizontal components of the momentum equation are now

$$\frac{\mathrm{d}u}{\mathrm{d}t} = -\frac{1}{\rho}\frac{\partial P}{\partial x} + fv, \tag{7.22}$$

$$\frac{\mathrm{d}v}{\mathrm{d}t} = -\frac{1}{\rho}\frac{\partial P}{\partial y} - fu, \tag{7.23}$$

where $u$ and $v$ are the $x$ and $y$ components of velocity, and where $x$ is east and $y$ is north. We have transformed to a local coordinate system where the curvature of the Earth's surface is ignored. Now, for sufficiently large-scale motions (or equivalently at low Rossby number) the acceleration terms on the left-hand sides of (7.22) and (7.23) can be replaced with zero; this is called the *geostrophic approximation*. In situations where this approximation applies, we can calculate the velocity field if we know the pressure field; we can see from these equations that the velocity is directed along the contours of pressure (isobars). This is the reason for the familiar patterns

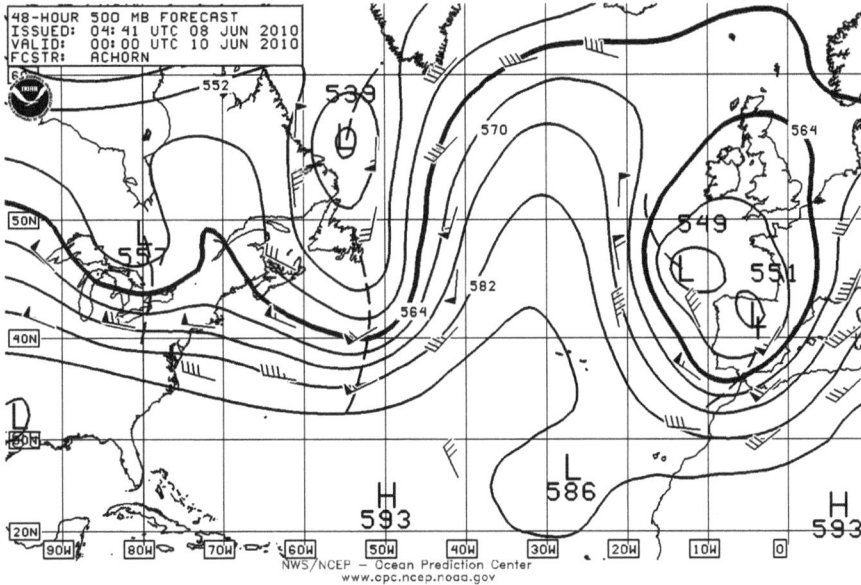

**Figure 7.7.** A weather forecast for 10 June 2010. Note that the wind is roughly parallel to the isobars.

of anticlockwise winds around a low-pressure area (cyclone) and clockwise winds around a high-pressure area (anticyclone)—see figure 7.7. In the southern hemisphere the directions are reversed along with the sign of $f$.

## 7.9 Rossby waves

We conclude this chapter on rotating fluids with a brief look at a wave which is ubiquitous in the Earth's atmosphere and ocean (Rossby 1939) and which is perhaps also interesting in astrophysics, for instance in fast-spinning neutron stars where it may be responsible for the emission of gravitational radiation.

First of all, let us assume that the Rossby number of a system is small: Ro $\ll 1$. We shall look at the purest kind of Rossby wave, namely that where the motion takes place on a plane, and where the motion on that plane is incompressible. Using the shallow water equations (5.17) and (5.18) together with the form of the Coriolis term as in (7.22) and (7.23), in the linear regime we have

$$\frac{\partial u}{\partial t} - fv = -g\frac{\partial \zeta}{\partial x}, \tag{7.24}$$

$$\frac{\partial v}{\partial t} + fu = -g\frac{\partial \zeta}{\partial y}, \tag{7.25}$$

$$\frac{\partial u}{\partial x} + \frac{\partial v}{\partial y} = 0, \tag{7.26}$$

where the Coriolis parameter $f$ now depends on latitudinal coordinate $y$. To be more precise $f = f_0 + \beta y$ where $f_0$ and $\beta$ are constants, which is valid as long as we are not looking at too large a region in latitude. This is called the *β-plane approximation*, as opposed to the *f-plane approximation* where $f$ is a constant. Cross differentiating and subtracting from one another the horizontal momentum equations, and assuming as usual a solution of the form $e^{i(kx+ly-\omega t)}$, we have

$$l\omega u - \beta v - flv - k\omega v - fku, \tag{7.27}$$

$$ku + lv = 0, \tag{7.28}$$

which can easily be combined to give the dispersion relation

$$\omega = -\frac{k\beta}{k^2 + l^2}. \tag{7.29}$$

The phase velocity, given in general in a system with more than one spatial dimension by $\omega \mathbf{k}/|\mathbf{k}|^2$, turns out to be equal to $-\omega^2 \mathbf{k}/(k\beta)$. The $x$-component of the phase velocity always has opposite sign to $\beta$, meaning that the waveform always moves in the opposite direction to the rotation of the system. On Earth, the waves move to the west. However, note that the group velocity can be in any direction, depending on the wavevector. Figure 7.8 shows a typical Rossby wave propagating around the Earth's northern hemisphere.

Alternatively we can take the curl of the momentum equation:

$$\nabla_h \times \frac{\partial \mathbf{u}_h}{\partial t} + \nabla_h \times ((\mathbf{u}_h \cdot \nabla_h)\mathbf{u}_h) = -\nabla_h \times (f\mathbf{e}_z \times \mathbf{u}_h), \tag{7.30}$$

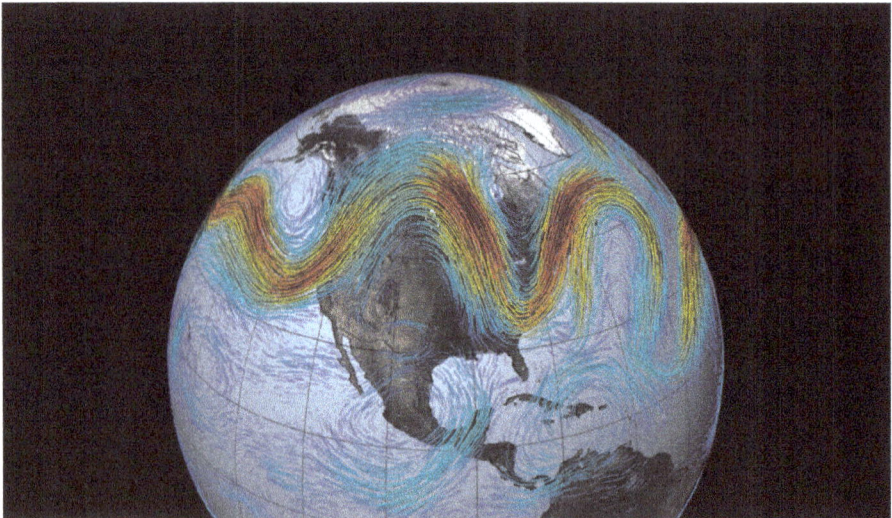

**Figure 7.8.** A Rossby wave propagating around the northern hemisphere. The velocity field is shown and colour-coded for magnitude. Note the jet stream at the boundary between warm and cold air. (Image credit: NASA/GSFC.)

where $\mathbf{e}_z$ is the vertical unit vector. In this two-dimensional velocity field $\mathbf{u}_h$ the vorticity is parallel to $\mathbf{e}_z$ and we shall call it $\psi$, a scalar. After a little algebra we arrive at the shallow-water vorticity equation

$$\frac{\mathrm{d}}{\mathrm{d}t}(f + \psi) = -(f + \psi)\nabla_h \cdot \mathbf{u}_h, \qquad (7.31)$$

which is analogous to (7.8). To obtain the simplest modes we could again take $\nabla_h \cdot \mathbf{u}_h = 0$, write out the Lagrangian derivative in its various parts and then assume a solution of the form $e^{i(kx+ly-\omega t)}$. However, the purpose of looking at the vorticity equation is to obtain an intuitive understanding of the wave. Since the absolute vorticity $f + \psi$ is conserved if $\nabla_h \cdot \mathbf{u}_h = 0$, if material which initially has relative vorticity $\psi = 0$ moves northwards or southwards (i.e. to a region of different $f$) it will start to rotate. Therefore any perturbation to the latitudinal velocity field will propagate to the west.

# Exercises

### 7.1 Conservation of angular momentum
Show that the two terms $-\omega(\nabla \cdot \mathbf{u})$ and $(\boldsymbol{\omega} \cdot \nabla)\mathbf{u}$ in (7.2) can be written together as $-\omega(\nabla_\perp \cdot \mathbf{u})$ where $\nabla_\perp$ is the divergence in a place perpendicular to $\boldsymbol{\omega}$. By considering a small rotating spheroidal fluid element of equatorial and polar (i.e. normal and parallel to $\boldsymbol{\omega}$) radius $a$ and $b$ in a velocity field, show that this represents the conservation of angular momentum as $a$ and $b$ change. Now take (7.2) and using the continuity equation turn it into an expression for the Lagrangian derivative of $\omega/\rho$.

### 7.2 Behaviour of smoke rings
Using the method of images, describe what happens when a vortex ring (e.g. smoke ring) approaches a wall. In addition, describe the interaction of two vortex rings, both of the same and opposite senses of rotation, where the line joining the centres of the two rings is perpendicular to the planes of the rings.

### 7.3 Tornadoes
A typical tornado can be approximated by a Rankine vortex with a core radius of 50 m, with a wind speed at this radius of 50 m s$^{-1}$. Write down an expression for the wind speed as a function of radius, and find an expression for the pressure as a function of radius. Assuming no external driving, find the speed and direction the tornado moves when it is a horizontal distance 200 m below a high cliff.

### 7.4 Rotational flattening of stars and planets
Find an approximate expression for the extent of the rotational flattening of a star as a function of rotation period, and show that there is a lower limit to the rotation period at which the centrifugal force is comparable to gravity. Write this limit in terms of the mean density of the star.

### 7.5 Water in a spinning tank
A tank of water is made to rotate at a constant angular velocity. Show that the surface of the water is a parabola shape.

### 7.6 Shallow-water vorticity equation

Show that it is possible to rewrite (7.31) as

$$\frac{d}{dt}\left(\frac{f + \psi}{\xi}\right) = 0, \tag{7.32}$$

where $\xi$ is now the total depth of the fluid including equilibrium and perturbation depths so that $\xi = h + \zeta$; in section 7.9 it was simply assumed that the Lagrangian derivative of $\xi$ was zero. Comment on the consequences of the conservation of this quantity in terms of mechanisms for generating relative vorticity $\psi$ in an ocean.

### 7.7 Bathtub

We saw in section 7.5 that in a rotating frame of reference, an initially non-rotating body of fluid can be brought into rotation simply by changing its cross-section in a plane perpendicular to the frame rotation $\Omega$. Do an order of magnitude calculation of the angular velocity expected as bathwater goes down the plughole, and comment on the result.

### 7.8 Vorticity in two and three dimensions

Only one of the four terms on the right-hand side of (7.8) has an equivalent in (7.31), namely the compression/expansion term. In the derivation of (7.31) we ignored viscosity; if we had not done so there would be a viscous term too. Why are the other two terms in (7.8) absent in (7.31)? Refer to section 7.2.

## References

Busse F H 1982 On the problem of stellar rotation *ApJ* **259** 759–66

Rossby C-G 1939 Relation between variations in the intensity of the zonal circulation of the atmosphere and the displacements of the semi-permanent centers of action *J. Marine Res.* **2** 38–55

Sweet P A 1950 The importance of rotation in stellar evolution *Mon. Not. R. Astron. Soc.* **110** 548

Zahn J-P 1992 Circulation and turbulence in rotating stars *Astron. Astrophys.* **265** 115–32

von Zeipel H 1924 The radiative equilibrium of a rotating system of gaseous masses *Mon. Not. R. Astron. Soc.* **84** 665–83

# Chapter 8

## Magnetohydrodynamics: equations and basic concepts

In this chapter the non-relativistic magnetohydrodynamics (MHD) equations are derived, starting with the non-magnetic fluid equations and then using Maxwell's equations to add the magnetohydrodynamic terms. Following that, some basic ideas in MHD are described, which are useful in building up an intuitive understanding of the subject. Students who are just using this book to learn about MHD might wish to refer to section 1.2 where the hydrodynamic equations are derived.

### 8.1 The MHD equations

The MHD equations are essentially an extension of the hydrodynamic equations with one extra variable: the magnetic field. There is one extra term in the momentum equation and a new partial differential equation called the induction equation. This is merely an extension to the equations, no more complicated than other extensions such as rotation or viscosity[1].

We shall temporarily abandon the fluid picture, going back to individual particles with which the student may be more familiar from previous courses. The force on a single particle of charge $q$ moving with velocity $\mathbf{v}$ in an electromagnetic field is given by:

$$\mathbf{F} = q\left(\mathbf{E} + \frac{\mathbf{v}}{c} \times \mathbf{B}\right), \tag{8.1}$$

where $\mathbf{E}$ and $\mathbf{B}$ are the electric and magnetic fields and $c$ is the speed of light[2]. This force is normally called the Lorentz force. Now, in MHD we are interested in the

---

[1] 'Magnetohydrodynamics' is a long word and the subject is perceived to be intrinsically difficult. Consequently, relatively few people have ever entered the field. As a result, it isn't a 'field' but a lush orchard with low-hanging fruit in joyous abundance!

[2] In some sense we can use this as the *definition* of $\mathbf{E}$ and $\mathbf{B}$, by saying that the force on a particle is some function of its charge and velocity which can be characterised by two vector fields in the form of (8.1).

doi:10.1088/978-1-6817-4597-8ch8          8-1

force on the fluid as a whole rather than on individual particles—the total force per unit volume (also called the Lorentz force, confusingly) is therefore

$$\mathbf{F}_{\text{Lor}} = \mathbf{F}_i + \mathbf{F}_e = (n_i q_i + n_e q_e)\mathbf{E} + \left(n_i q_i \frac{\bar{\mathbf{v}}_i}{c} + n_e q_e \frac{\bar{\mathbf{v}}_e}{c}\right) \times \mathbf{B}, \tag{8.2}$$

where $\mathbf{F}_i$ is the total force on the ions, $n_i$, $q_i$ and $\bar{\mathbf{v}}_i$ are the number density, charge and mean velocity of ions, and the quantities with subscript e refer to electrons. We have assumed here that all ions have the same charge and that there are no neutral particles, but it is trivial to show that a generalisation does not affect the end result. We now define the fractional charge imbalance ratio $\varepsilon \equiv (n_i q_i + n_e q_e)/n_e q_e$, as well as the drift velocity as the mean velocity of the electrons relative to the fluid, i.e. $\mathbf{v}_{\text{drift}} \equiv \bar{\mathbf{v}}_e - \mathbf{u}$, where the fluid velocity $\mathbf{u} \approx \bar{\mathbf{v}}_i$ since the ions carry almost all of the momentum. Let us now rewrite (8.2) as

$$\mathbf{F}_{\text{Lor}} = n_e q_e \left[\varepsilon \mathbf{E} + \left(\varepsilon \frac{\mathbf{u}}{c} + \frac{\mathbf{v}_{\text{drift}}}{c}\right) \times \mathbf{B}\right]. \tag{8.3}$$

Noting that the Earth needs $\varepsilon < 10^{-36}$ so that the electric field does not overcome gravity and cause it to explode and that in almost all astrophysical contexts $\varepsilon$ is negligible, we drop terms with $\varepsilon$ (despite the fact that normally $\mathbf{v}_{\text{drift}} \ll \mathbf{u}$; more on this in section 8.2). It is now convenient to introduce the concept of electric current density $\mathbf{J} = n_e q_e \mathbf{v}_{\text{drift}}$, with which we simplify the Lorentz force (a force per unit volume) to

$$\mathbf{F}_{\text{Lor}} = \frac{1}{c}\mathbf{J} \times \mathbf{B}. \tag{8.4}$$

We still need to answer the question of the origin of this relative velocity of the electrons to the ions; this comes essentially from the difference in force on the two species. The electrons experience a force relative to the fluid given by

$$\mathbf{F}_e - \mathbf{F}_{\text{Lor}} = n_e q_e \left(\mathbf{E} + \frac{\mathbf{u}}{c} \times \mathbf{B}\right), \tag{8.5}$$

where $\mathbf{v}_{\text{drift}} \ll \mathbf{u}$ was used. This force will accelerate the electrons relative to the ions, and there will quickly be a balance established between this acceleration and losses through collisions between electrons and ions. In fluids with normal conductivity properties, the drift velocity (and therefore current) established will be proportional to this acceleration. This gives us Ohm's law:

$$\mathbf{J} = \sigma \left(\mathbf{E} + \frac{\mathbf{u}}{c} \times \mathbf{B}\right), \tag{8.6}$$

where $\sigma$ is the conductivity of the fluid, which will depend on mean free path, temperature, etc.

So far, we have added a term (8.4) to the momentum equation containing two new variables $\mathbf{B}$ and $\mathbf{J}$. Ohm's law (8.6) introduces yet another new variable $\mathbf{E}$, so that we need an additional two equations to close the set. Maxwell's equations are

$$\nabla \cdot \mathbf{E} = 4\pi\rho_e, \tag{8.7}$$

$$\nabla \cdot \mathbf{B} = 0, \tag{8.8}$$

$$\nabla \times \mathbf{E} = -\frac{1}{c}\frac{\partial \mathbf{B}}{\partial t}, \tag{8.9}$$

$$\nabla \times \mathbf{B} = \frac{1}{c}\frac{\partial \mathbf{E}}{\partial t} + \frac{4\pi}{c}\mathbf{J}, \tag{8.10}$$

where $\rho_e$ is the net charge density[3]. First of all, note that if (8.8) is satisfied at some point in time, (8.9) ensures that it is satisfied at all other times, since the divergence of the curl of any vector field is zero. In standard MHD we now make the approximation that the charge density $\rho_e$ is small; also that the displacement current in (8.10) can be neglected, i.e. that $4\pi\mathbf{J} \gg \partial\mathbf{E}/\partial t$. (See section 8.2 for a justification.)

From Ohm's law (8.6) we obtain an expression for the electric field $\mathbf{E} = (1/\sigma)\mathbf{J} - (\mathbf{u}/c) \times \mathbf{B}$, which we can use in conjunction with (8.9) to obtain

$$\frac{\partial \mathbf{B}}{\partial t} = -c\nabla \times \mathbf{E} = \nabla \times \left(\mathbf{u} \times \mathbf{B} - \frac{c}{\sigma}\mathbf{J}\right), \tag{8.11}$$

and dropping the displacement current from (8.10) and substituting for $\mathbf{J}$ gives the induction equation

$$\frac{\partial \mathbf{B}}{\partial t} = \nabla \times (\mathbf{u} \times \mathbf{B} - \eta\nabla \times \mathbf{B}), \tag{8.12}$$

where magnetic diffusivity has been defined $\eta \equiv c^2/4\pi\sigma$, with units cm s$^{-1}$. Finally we can substitute for $\mathbf{J}$ into the Lorentz force (8.4) to give force per unit volume

$$\mathbf{F}_{\text{Lor}} = \frac{1}{4\pi}(\nabla \times \mathbf{B}) \times \mathbf{B}. \tag{8.13}$$

Thus $\mathbf{E}$ and $\mathbf{J}$ have been eliminated. In summary, compared to the original hydrodynamics equations we have one additional variable $\mathbf{B}$, one additional equation (8.12) and one additional term (8.13) in the momentum equation. Note that nothing changes if we reverse the direction of the magnetic field—$\mathbf{B}$ is a so-called 'pseudo-vector'.

---

[3] Note the similarity between (8.7) and the equation relating the gravitational field to density $\nabla \cdot \mathbf{g} = -4\pi G\rho$. No constant is required in the electromagnetic equivalent because it is built into the unit of charge. The other difference of course is that $\rho_e$ can be either positive or negative.

## 8.2 The MHD approximation

In addition to the standard conditions under which the fluid approximation is valid, e.g. collision frequency, etc, we have made further approximations. We assumed that the magnetic permeability and dielectric permittivity of the plasma can be ignored (refractive index equal to unity). We also assumed the flow is non-relativistic. Finally we made what is known as the 'MHD approximation', according to which the conductivity of the material is high enough so that the charge density $\rho_e$ is low and (8.7) can be ignored.

In the rest frame of a test particle (in which quantities are denoted with a prime), the force experienced is $\mathbf{F}' = q\mathbf{E}'$. In transforming to an inertial frame, we ignore terms in $v^2/c^2$ (so that the Lorentz factor $\Gamma \approx 1$) to obtain the lab frame relation (8.1). Returning to the fluid picture, a high conductivity $\sigma$ ensures that current can flow in order to almost neutralise the rest frame electric field $\mathbf{E}' = \mathbf{E} + (\mathbf{u}/c) \times \mathbf{B}$ so that $E' \ll E$ and $E \sim (u/c)B$. We can now justify neglecting the displacement current in (8.10), because it is of order $E(u/c)/L \sim B(u/c)^2/L$ where $L$ is a typical length scale of the flow. This is smaller by a factor $v^2/c^2$ than the curl of the magnetic field; meaning that $J \sim cB/L$ (dropping factors of $4\pi$).

Looking at the first of Maxwell's equations (8.7), also known as Gauss' law, we see that $\rho_e \sim E/L \sim (u/c)B/L \sim (u/c)J/c$. Since this law holds in every reference frame it follows from $E' \ll E$ that $\rho_e' \ll \rho_e$. Current density $\mathbf{J}$ can be considered equal in lab and co-moving frames since assuming that $\Gamma \approx 1$ gives $\mathbf{J}' = \mathbf{J} - \rho_e\mathbf{u}$; the ratio of the two terms is $(u/c)^2$ so that $\mathbf{J}' = \mathbf{J}$. Magnetic field can also be considered frame-independent, since a transformation assuming $\Gamma \approx 1$ gives $\mathbf{B}' = \mathbf{B} - (\mathbf{u}/c) \times \mathbf{E}$, and the electric field $\mathbf{E}$ is itself smaller than the magnetic field by a factor $u/c$ so we are left with a ratio $u^2/c^2$ between the two terms so that $\mathbf{B} = \mathbf{B}'$.

There are obviously astrophysical contexts in which these approximations do not hold, for instance relativistic flows such as gamma-ray burst jets, or situations where plasma effects become important or where the fluid approximation breaks down; these are outside the scope of this book.

## 8.3 The magnetic and other fields

The MHD equations in the form of (8.12) and (8.13) contain magnetic fields but not current density or electric fields. In fact there are only a few contexts in astrophysics where it is necessary to think about these extra fields. In the comoving frame, i.e. the frame 'felt' by the fluid, we have seen that the electric field vanishes in the case of infinite conductivity and in other, non-relativistic contexts it is very small compared to the magnetic field. This, combined with the fact that the fluid is normally almost perfectly neutral, renders its effect totally insignificant, except in special contexts where densities are very low and velocities are relativistic, such as the magnetospheres of neutron stars. Likewise, we need only think about current density in such special situations.

It is therefore surprising that the literature on MHD often refers to these additional fields in context where the MHD approximation can be considered to hold perfectly. The reason for this can historically be traced back to a transfer of understanding of terrestrial phenomena, particularly that of electric circuits, coils, inductances and so on to the astrophysical context. However, the two contexts are rather different. For example, consider using a battery to pass a current through a wire and measuring the magnetic field it produces in the insulating fluid (air) surrounding it. None of these exists in astrophysics—there are no batteries, no wires, and almost no insulating fluids. It makes no sense in MHD to say that a magnetic field is produced by a current—the two are related by $(4\pi/c)\mathbf{J} = \nabla \times \mathbf{B}$ and that is the end of the story. In fact, the magnetic field can in some sense be considered the more fundamental of the two, as its evolution is governed by conservation laws which do not apply to the current. The student is advised to avoid thinking in terms of currents, electromotive forces and circuits as these will distract from an understanding of the subject; only $\mathbf{B}$ is required!

## 8.4 A brief note concerning units

At this juncture it is worth commenting on the difference between the c.g.s. units employed here and the SI units often taught in undergraduate courses. The reader will notice that the equations above contain only one constant of nature: the speed of light $c$. In contrast, a glance at some of the text books reveals that the SI system is burdened not only with $c$ but also with the rather 19th century concepts of the permittivity and permeability of the ether, $\varepsilon_0$ and $\mu_0$. Another advantage of c.g.s. is that electric and magnetic fields have the same units. However, a word of caution: there exist variations of c.g.s. units; here we use 'Gaussian c.g.s.' units, which are fairly standard in astrophysics. In this system the unit of charge, called the statcoulomb, is defined from Coulomb's law of the force between two point charges $F = q_1q_2/r^2$ such that two unit charges at a separation of 1 cm experience a repulsion of 1 dyne. It can be written in terms of the other units: 1 statC = 1 $g^{1/2}$ $cm^{3/2}$ $s^{-1}$. Occasionally in astrophysics one comes across Heaviside–Lorentz units which differ from Gaussian only in factors of $4\pi$. In SI units the unit of charge is defined differently and Coulomb's law requires a constant $1/4\pi\varepsilon_0$.

The unit of magnetic field in the c.g.s. system is the gauss (abbreviation G), which is equal to $10^{-4}$ tesla, the unit in the SI system. Magnetic fields strengths in the Universe range from $10^{-6}$ G in the intergalactic medium, 0.3–0.6 G on the surface of the Earth, $10^4$ G on some magnetic main-sequence stars, up to $10^6$ G in laboratories, up to $10^9$ G in white dwarfs and up to $10^{15}$ G in neutron stars.

## 8.5 Field lines, flux conservation and flux freezing

We assume for most of the rest of the chapter that we have infinite conductivity (i.e. $\eta = 0$), which taking $\eta = 0$ gives the ideal MHD induction equation

$$\frac{\partial \mathbf{B}}{\partial t} = \nabla \times (\mathbf{u} \times \mathbf{B}). \tag{8.14}$$

Let us define a magnetic flux $\phi$ as an integral of the normal component of $\mathbf{B}$ on a surface S

$$\phi = \int_S \mathbf{B} \cdot d\mathbf{S} \tag{8.15}$$

and then calculate the change of flux through that surface as it moves with the flow of the fluid:

$$\frac{d\phi}{dt} = \int_S \frac{\partial \mathbf{B}}{\partial t} \cdot d\mathbf{S} + \oint_l (\mathbf{u} \times d\mathbf{l}) \cdot \mathbf{B}, \tag{8.16}$$

where the first term comes from the rate of change of flux through the surface if it were fixed in space and the second comes from the movement of the surface from the fluid velocity $\mathbf{u}$. The surface is bounded by a line $l$. Substituting (8.14) into this equation and using Stokes' theorem, the first term becomes $\oint_l \mathbf{u} \times \mathbf{B} \cdot d\mathbf{l}$. From the triple vector product rule we now see that the two terms cancel and that the flux through the co-moving surface is constant in time.

If we now imagine the fluid being composed of small co-moving fluid elements, each threaded by a constant flux, it becomes clear that the concept of field lines and of their being 'frozen' into the fluid are useful tools in understanding MHD. This will be discussed below at greater length.

Note that this result is analogous to that of freezing of vorticity in hydrodynamics. The vorticity equation (7.2) in an inviscid barotropic flow with conservative body forces can be written

$$\frac{\partial \omega}{\partial t} = \nabla \times (\mathbf{u} \times \omega), \tag{8.17}$$

which has the same form as (8.14).

## 8.6 Magnetic diffusivity

If we assume that the electrical conductivity $\sigma$ is uniform, we can rearrange the induction equation (8.12) using the constraint $\nabla \cdot \mathbf{B} = 0$ and the vector identity $\nabla \times \nabla \times \mathbf{A} = \mathbf{A}(\nabla \cdot \mathbf{A}) - \nabla^2 \mathbf{A}$, to

$$\frac{\partial \mathbf{B}}{\partial t} = \nabla \times (\mathbf{u} \times \mathbf{B}) + \eta \nabla^2 \mathbf{B}, \tag{8.18}$$

where the magnetic diffusivity $\eta$, like the kinetic and thermal diffusivities $\nu$ and $\chi$, has units cm$^2$ s$^{-1}$. We give names to the ratios of diffusivities: the Prandtl number and magnetic Prandtl number are defined as $\mathrm{Pr} \equiv \nu/\chi$ and $\mathrm{Pr_m} \equiv \nu/\eta$ respectively. Note the similarity with the diffusive terms in the momentum and heat equations—all contain a diffusive coefficient multiplied by $\nabla^2$ of the relevant variable. As with the other diffusivities, from analysis of units we see that there is a characteristic magnetic diffusive timescale, equal to $L^2/\eta$ where $L$ is the characteristic length scale of the system. This is called the *Ohmic timescale*. In many contexts the Ohmic timescale is very much longer than other timescales of interest and it is possible to ignore the diffusive term in the inductive equation. This regime is called *ideal MHD*. In many, if not most, applications we can use ideal MHD.

Finally, it is worth mentioning the heating from Ohmic dissipation, known as *Joule heating*: per unit volume this is equal to $\mathbf{J} \cdot \mathbf{E}'$ where $\mathbf{E}'$ is the electric field in the comoving frame (recalling that current density is the same in both frames). Expressed differently, this means that

$$\rho \, q_{\text{Joule}} = \frac{1}{\sigma} J^2 = \frac{\eta}{4\pi} (\nabla \times \mathbf{B})^2, \tag{8.19}$$

where $q$ is the heating per unit mass (not volume, in keeping with previous chapters), hence the factor of $\rho$.

## 8.7 Magnetic pressure, tension and energy density

The Lorentz force can be written in an alternative form, making use of a vector identity and the solenoidal constraint $\nabla \cdot \mathbf{B} = 0$:

$$\mathbf{F}_{\text{Lor}} = \frac{1}{4\pi} (\nabla \times \mathbf{B}) \times \mathbf{B} = -\nabla \left( \frac{B^2}{8\pi} \right) + \frac{1}{4\pi} (\mathbf{B} \cdot \nabla) \mathbf{B}. \tag{8.20}$$

The first term looks like the pressure gradient term $-\nabla P$, so the quantity $B^2/8\pi$ is called the *magnetic pressure*. The second term on the right is often called the magnetic tension or curvature force. However, we need to remind ourselves that the total Lorentz force is always perpendicular to the magnetic field, so that the components of these two terms parallel to the field must cancel. After some manipulation we can rewrite the Lorentz force without these cancelling components as

$$\mathbf{F}_{\text{Lor}} = -\nabla_{\perp} \frac{B^2}{8\pi} + \kappa \frac{B^2}{4\pi}, \tag{8.21}$$

where $\kappa = [(\mathbf{B}/B) \cdot \nabla](\mathbf{B}/B)$ is the curvature vector (directed along the radius of curvature of the field and equal in magnitude to the reciprocal of that radius) and $\nabla_{\perp}$ is the part of the gradient perpendicular to the field. The curvature force resembles the tension in a string in that it will tend to restore a perturbed straight field line to its original shape.

It can be shown that the magnetic pressure $B^2/8\pi$ is also the energy density of the magnetic field; to demonstrate this in an MHD context in a non-rigorous way is reasonably straightforward. Imagine a straight tube of infinite length with cylindrical cross-section of radius $a$ (figure 8.1). It contains a uniform field $B$ parallel to its length and is surrounded by unmagnetised fluid. The curvature force vanishes everywhere and the only place where the other part of the Lorentz force does not vanish is the boundary, where it is a delta function directed normal to the boundary. Assuming equilibrium, this must be balanced by a delta function in the gas pressure gradient force, i.e. a discontinuity in pressure, where the gas pressure outside the tube is greater than that inside by a quantity $B^2/8\pi$. Now imagine making an adiabatic change in $a$ so that the flux $\phi = \pi a^2 B = \text{const}$. If the magnetic field has energy $e$ per unit volume, then the magnetic energy of the tube is $E = \pi a^2 e$ per unit length. From the $dU = -p \, dV$ relation in thermodynamics (doing work

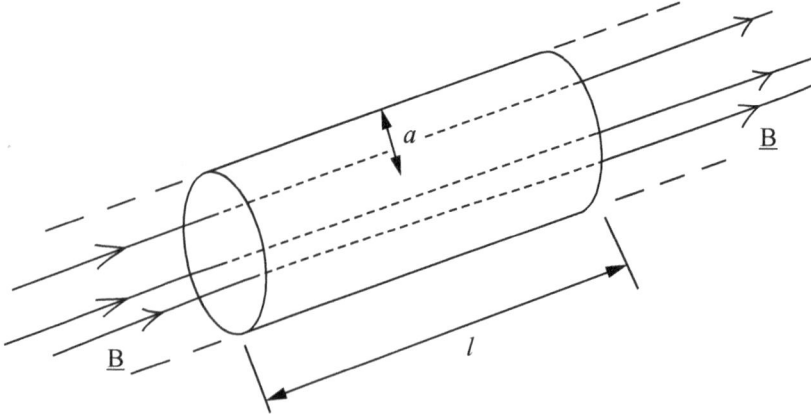

**Figure 8.1.** A section of a flux tube.

on the magnetic field by pushing against the Lorentz force) we therefore have $dE = -(B^2/8\pi)2\pi a \, da = -\phi^2/(4\pi^2 a^3)da$ and so assuming that the energy $E$ goes to zero at $a = \infty$ we can integrate from $a' = \infty$ to $a$ to give

$$\int_0^E dE' = -\int_\infty^a \frac{\phi^2}{4\pi^2 a'^3} da' \implies E = \frac{\phi^2}{8\pi^2 a^2} = \frac{B^2}{8\pi}\pi a^2 \implies e = \frac{B^2}{8\pi}. \quad (8.22)$$

Magnetic pressure is different from gas pressure in that it is not isotropic. To demonstrate this, imagine a section of the aforementioned flux tube of length $l$, which contains magnetic energy $E = \pi a^2 l B^2/8\pi$. Stretching this section of the tube while keeping the cross-section $a$ fixed will not change $B$ so the increase in energy is simply $dE = \pi a^2 dl B^2/8\pi$. Equating this to the energy $dE = -P \, dV$ again, we have $-P_{mag}\pi a^2 dl = (B^2/8\pi)\pi a^2 dl$ and so the tube has a tension per unit area of $B^2/8\pi$, which can be thought of as a negative pressure. Of course, no infinite tube exists in reality—however, we can imagine a tube being connected to itself in a loop, which helps us to understand how the tension comes about even though the Lorentz force is always perpendicular to the field.

If we repeat the above thought experiment with not a stretch or compression in one direction, but an *isotropic* expansion or compression of a magnetised fluid element, it is easily shown that the resistance of the magnetic field to such a compression is equivalent to an isotropic pressure of $B^2/24\pi$. This is simply the average of a pressure of $B^2/8\pi$ in the two directions perpendicular to the field and tension of $B^2/8\pi$ parallel to it. Furthermore, it can be shown that any self-contained magnetic feature (i.e. one without field lines crossing its boundary) exerts a mean pressure $B^2/24\pi$ on its surroundings, meaning that in equilibrium, the gas pressure outside the feature must be greater than the average pressure inside by this quantity. This should not be surprising since magnetic field is a relativistic phenomenon and relativistic fluids (e.g. photons, relativistic particles, gravity) exert a pressure equal to one third of their energy density—equivalent to an adiabatic index of 4/3. However, one must always be careful when simplifying the effect of a magnetic field to an isotropic magnetic pressure since the geometry of the field is often crucial.

## 8.8 Waves

The restoring force from bending field lines and from squeezing field lines together allows the propagation of waves in a magnetised fluid. The simplest form of waves propagates in a medium of initially constant pressure and density threaded by a uniform magnetic field, and since magnetic pressure is not isotropic it is necessary in the general case to consider the angle between the wavevector and the magnetic field; however the equivalence of the two dimensions perpendicular to the field allows us to drop one and consider only the remaining two dimensions. It turns out that there are three kinds of wave in a compressible magnetised fluid: the Alfvén wave and the fast and slow magnetoacoustic (or magnetosonic) waves. First of all we write the linearised momentum, continuity, induction and energy equations:

$$\frac{\partial \mathbf{u}}{\partial t} = -\frac{1}{\rho}\nabla \delta P + \frac{1}{4\pi\rho}(\nabla \times \delta \mathbf{B}) \times \mathbf{B}, \tag{8.23}$$

$$\frac{\partial \delta \rho}{\partial t} = -\rho \nabla \cdot \mathbf{u}, \tag{8.24}$$

$$\frac{\partial \delta \mathbf{B}}{\partial t} = \nabla \times (\mathbf{u} \times \mathbf{B}), \tag{8.25}$$

$$\delta P = c_{\mathrm{s}}^2 \delta \rho, \tag{8.26}$$

where $\mathbf{B}$ is the equilibrium field which is parallel to the $y$-axis and quantities with $\delta$ are the perturbations which, as well as the $x$ and $y$ components of the velocity field $u$ and $v$, are small. Writing $\tilde{\rho} = \delta\rho/\rho$ and $\mathbf{b} = \delta\mathbf{B}/B$ and substituting from (8.26) we can write out the equations as

$$\frac{\partial u}{\partial t} = -c_{\mathrm{s}}^2 \frac{\partial \tilde{\rho}}{\partial x} + v_{\mathrm{A}}^2 \left(\frac{\partial b_x}{\partial y} - \frac{\partial b_y}{\partial x}\right), \tag{8.27}$$

$$\frac{\partial v}{\partial t} = -c_{\mathrm{s}}^2 \frac{\partial \tilde{\rho}}{\partial y}, \tag{8.28}$$

$$\frac{\partial \tilde{\rho}}{\partial t} = -\frac{\partial u}{\partial x} - \frac{\partial v}{\partial y}, \tag{8.29}$$

$$\frac{\partial b_x}{\partial t} = \frac{\partial u}{\partial y}, \tag{8.30}$$

$$\frac{\partial b_y}{\partial t} = -\frac{\partial u}{\partial x}, \tag{8.31}$$

where the Alfvén speed has been defined as $v_A \equiv B/\sqrt{4\pi\rho}$. To work through the derivation of the dispersion relation is a little lengthy so we restrict ourselves here to two special cases, where the wavevector is parallel and perpendicular to the field. In the parallel case $\partial/\partial x = 0$ and the equations split into two sets, one set made of (8.28) and (8.29) containing $v$ and $\tilde{\rho}$ and the other set made of (8.27) and (8.30), containing just $u$ and $b_x$. The first set describes a longitudinal wave, since the motion is parallel to the wavevector. Motion parallel to a magnetic field has no effect on it (since $\mathbf{u} \times \mathbf{B}$ present in the induction equation vanishes), and this wave is simply a sound wave. The other set describes a transverse wave called an *Alfvén wave* which is non-compressional (since $\tilde{\rho}$ is absent). It is left as an exercise for the student to derive the dispersion relation for these waves, and to show that the propagation speed is $v_A$.

The other special case is that of perpendicular wavevector and field, where $\partial/\partial y = 0$. Here, $v$ and $b_x$ drop out of the equations and there is only one wave, a longitudinal wave called the *fast magnetoacoustic wave*. It is left as an exercise to derive the dispersion relation and to show that the propagation speed is $(c_s^2 + v_A^2)^{1/2}$. These waves are similar to sound waves, but the magnetic field provides an extra restoring force.

If the wavevector and field are neither parallel nor perpendicular, two waves are possible (the *fast* and *slow magnetoacoustic waves*), and the phase and group velocities are no longer parallel. These waves, which are not easy to visualise, are compressional and involve a mixture of gas pressure and magnetic restoring forces. Fortunately it is rarely necessary to know the detail.

Finally, a particular kind of Alfvén wave deserves a mention: the *torsional Alfvén wave*. Imagine a flux tube which is perturbed not by a sideways motion but by a twisting motion—the twist propagates along the tube[4]. This kind of wave is very important in various astrophysical situations, for instance in differentially rotating stars and in star formation, since it carries angular momentum. Torsional and planar Alfvén waves are illustrated in figure 8.2.

## 8.9 Different regimes in MHD

We aim here to analyse the relative sizes of the various terms in the momentum and induction equations:

$$\rho\frac{d\mathbf{u}}{dt} = -\nabla P + \frac{1}{4\pi}(\nabla \times \mathbf{B}) \times \mathbf{B} + \rho\nu\nabla^2\mathbf{u}, \tag{8.32}$$

$$\frac{\partial \mathbf{B}}{\partial t} = \nabla \times (\mathbf{u} \times \mathbf{B}) + \eta\nabla^2\mathbf{B}. \tag{8.33}$$

Assigning typical flow parameters $L, U$ and $T = L/U$—typical length scale, velocity and timescale—and comparing the size of various terms in the equations (dropping

---

[4] According to MHD folklore Alfvén got the idea from growing sunflowers in the Swedish arctic, where the flowers rotate once every day during the summer to follow the Sun; at the end of the season the Sun finally sets and the sunflowers 'unwind'.

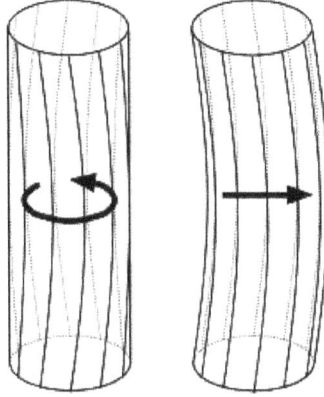

**Figure 8.2.** Torsional and plane Alfvén waves propagating along a flux tube. (Reproduced with permission from Van Doorsselaere *et al* 2008).

factors of order unity) we can rewrite these two equations in terms of non-dimensional variables and gradients such as $\mathbf{u}' = U\mathbf{u}$ (where we immediately drop the prime below) as:

$$\frac{d\mathbf{u}}{dt} = -\frac{c_s^2}{U^2}\nabla P + \frac{v_A^2}{U^2}(\nabla \times \mathbf{B}) \times \mathbf{B} + \frac{1}{\mathrm{Re}}\nabla^2\mathbf{u},$$

$$= \frac{1}{M^2}\left[-\nabla P + \frac{1}{\beta}(\nabla \times \mathbf{B}) \times \mathbf{B}\right] + \frac{1}{\mathrm{Re}}\nabla^2\mathbf{u}, \tag{8.34}$$

$$\frac{\partial\mathbf{B}}{\partial t} = \nabla \times (\mathbf{u} \times \mathbf{B}) + \frac{1}{\mathrm{Re_m}}\nabla^2\mathbf{B}, \tag{8.35}$$

where $c_s^2 = (\partial P/\partial\rho)_s$ and $v_A^2 = B^2/4\pi\rho$ are the sound and Alfvén speeds. Also included above is the familiar Reynolds number $\mathrm{Re} \equiv UL/\nu$ (section 4.4) which is now joined by the *magnetic Reynolds number* $\mathrm{Re_m} \equiv UL/\eta$. They are measures of the ratio of inertia to diffusivity of the two kinds. Also introduced is the so-called *plasma-$\beta$*, defined as $\beta \equiv 8\pi P/B^2 \approx c_s^2/v_A^2$. It is a (likely) ratio of the first and second terms on the right-hand side of the momentum equation. As we saw in previous chapters the Mach number is the ratio of flow speed to sound speed $M \equiv U/c_s$; sometimes one also speaks of the *Alfvénic Mach number* $M_A \equiv U/v_A$. Note that $\mathrm{Re_m}/\mathrm{Re} = \mathrm{Pr_m}$.

In an unmagnetised fluid, we can describe the flow with $M$ and Re (and possibly also the Strouhal number; see section 4.4). We have already seen that if $M \ll 1$ the flow is roughly incompressible, i.e. that $d\rho/dt \approx 0$, which gives the simplified continuity equation $\nabla \cdot \mathbf{u} = 0$. This is the regime we assumed in simplifying the viscous force in the momentum equation. The Reynolds number characterises the importance of viscosity.

In a magnetised medium we now have two extra parameters $\mathrm{Re_m}$ and $\beta$; the former describes the importance of diffusivity (finite conductivity) and the latter describes the relative importance of gas and magnetic pressures. The value of the 'plasma $\beta$' is very important in MHD. If $\beta \ll 1$, one expects from looking at the momentum equation that the Lorentz force will be much larger than the pressure gradient force. This means that velocities of order the Alfvén speed (which can be very high) will result unless the current $\nabla \times \mathbf{B}$ is almost parallel to $\mathbf{B}$, the so-called 'force-free' regime. Conversely, if $\beta \gg 1$ then the magnetic field will only have much effect on the flow in directions where the Lorentz force is not opposed by the pressure gradient or other stronger forces such as gravity.

## 8.10 Magnetic helicity

In section 8.5 we saw that the magnetic flux through a co-moving fluid surface is constant in the limit of high conductivity. We can now look at an additional quantity which is also conserved in this limit. Magnetic helicity is a global quantity defined as

$$H \equiv \int_V \mathbf{A} \cdot \mathbf{B} \, dV, \tag{8.36}$$

where $\mathbf{A}$ is the vector potential defined from $\mathbf{B} = \nabla \times \mathbf{A}$. Now, since the curl of the divergence of a scalar is zero, we can add any gradient of a scalar $\nabla \phi$ to the vector potential without changing the magnetic field; however this will in general affect the magnetic helicity. It can be shown in the following way (Woltjer 1958) though that magnetic helicity is gauge invariant provided that no magnetic field lines pass through the boundary of the volume of integration. Consider some new vector potential $\mathbf{A}' = \mathbf{A} + \nabla \phi$. The helicity is now

$$H' = \int_V [\mathbf{A} \cdot \mathbf{B} + (\nabla \phi) \cdot \mathbf{B}] \, dV \tag{8.37}$$

$$= H + \int_V [\nabla \cdot (\phi \mathbf{B}) - \phi (\nabla \cdot \mathbf{B})] \, dV \tag{8.38}$$

$$= H + \oint_S \phi \, \mathbf{B} \cdot d\mathbf{S}, \tag{8.39}$$

since the first term on the first line is simply equal to the original helicity; the second term was expanded with a standard vector identity to give the expression on the second line. Since the magnetic field is solenoidal (i.e. its divergence is zero), the second of the new terms vanishes, and the first can be rewritten with the aid of Gauss' theorem to give a surface integral. Therefore if $\mathbf{B} \cdot d\mathbf{S} = 0$ everywhere on the boundary, helicity is gauge invariant.

Helicity is a useful concept because of its conservation properties. It can be shown that it is perfectly conserved in the limit of infinite conductivity. First note that by integrating the ideal MHD induction equation (8.14) we have $\partial \mathbf{A}/\partial t = \mathbf{u} \times \mathbf{B}$. Now

$$\frac{\partial H}{\partial t} = \int_V dV \, [\mathbf{A}_t \cdot \mathbf{B} + \mathbf{A} \cdot \mathbf{B}_t] = \int_V dV \, [\mathbf{u} \times \mathbf{B} \cdot \mathbf{B} + \mathbf{A} \cdot \nabla \times (\mathbf{u} \times \mathbf{B})]$$

$$= \int_V dV [(\mathbf{u} \times \mathbf{B}) \cdot \nabla \times \mathbf{A} - \nabla \cdot (\mathbf{A} \times (\mathbf{u} \times \mathbf{B}))] \tag{8.40}$$

$$= -\oint_S d\mathbf{S} \cdot \mathbf{A} \times (\mathbf{u} \times \mathbf{B}).$$

It is sufficient then that the velocity goes to zero on the boundary of the domain.

In a fluid with high but finite conductivity, helicity is still approximately conserved—we can see this from a consideration of units. Now, the diffusive timescale on which the magnetic field decays due to finite conductivity, as we saw above, is $\tau \sim L^2/\eta$. This is shorter on shorter length scales so that when magnetic energy is converted to heat via Ohmic dissipation, it is mainly the small-scale structure where the energy is converted. Helicity however has units of length times energy and is therefore present more in the large scale components of the magnetic field than the magnetic energy, and less is therefore lost due to diffusive processes on small scales. Often then during MHD processes where the flow contains a range of length scales, energy is lost at the smallest scales while helicity is roughly conserved. This has been conserved by numerical experiments (Braithwaite 2015 and references therein).

Helicity also has units of flux squared, and can in fact be thought of in some sense of the product of two fluxes of different components of the magnetic field. It is often said that helicity is a measure of the 'twist' of the magnetic field, because a twisted field must contain at least two components. As we have seen, that twist is conserved even as magnetic energy is lost.

## 8.11 MHD equilibria

In many astrophysical contexts, we are interested in equilibrium situations where the forces are balanced. However, before we proceed, it is important to clarify what we mean by equilibrium. If we simply set the velocity to zero, the momentum equation (8.32) gives us a relation between $P$, $\mathbf{B}$ and $\rho\mathbf{g}$, so that any combination of the three which satisfies that relation will be an equilibrium. Also both sides of the continuity equation (1.9) go to zero, but the heat equation and the induction equation both contain diffusion terms so that the magnetic field and pressure field will evolve, giving rise to a non-zero velocity field. A truly stationary state is in general achievable only where thermal and magnetic diffusivities are both zero, $\chi = \eta = 0$. However, provided that these diffusivities are small, we can still find a *dynamic equilibrium* by setting $\mathbf{u} = \mathbf{0}$ and balancing forces. This equilibrium will not change appreciably on a dynamic timescale, i.e. the time taken for a sound or Alfvén wave to travel across the domain; rather, it will evolve over a longer timescale due to the diffusive terms.

So, finding a (dynamic) equilibrium is simply a matter of finding a solution to the following equation:

$$-\nabla P + \frac{1}{4\pi}(\nabla \times \mathbf{B}) \times \mathbf{B} + \rho\mathbf{g} = \mathbf{0}, \tag{8.41}$$

together with the constraint $\nabla \cdot \mathbf{B} = 0$. It is interesting to explore some of the properties of this equation. First of all, note that in the non-magnetised case it reduces to the equation of hydrostatic equilibrium $\nabla P = \rho\mathbf{g}$ which one often sees in the form $\partial P/\partial z = -\rho g$ where gravity is directed downwards along the $z$-axis. In the magnetised case, taking the dot product with $\hat{\mathbf{B}}$, the unit vector in the direction of the magnetic field, gives

$$(\hat{\mathbf{B}} \cdot \nabla)P = \rho(\mathbf{g} \cdot \hat{\mathbf{B}}) \qquad \text{or} \qquad \frac{\mathrm{d}P}{\mathrm{d}s} = \rho g_s, \tag{8.42}$$

where $\mathrm{d}P/\mathrm{d}s$ is the derivative along a field line and $g_s$ is the component of gravity along the field line. In other words, in an MHD equilibrium there is *hydrostatic balance along field lines*.

In a situation where $\beta \gg 1$ and the Lorentz term in (8.41) is much smaller than the pressure and gravity terms, we can imagine first constructing a non-magnetic equilibrium where $\nabla P = \rho\mathbf{g}$, adding a weak magnetic field and then making small adjustments to the pressure and density fields to balance the Lorentz force. In principle this should be possible, because an arbitrary magnetic field and its associated Lorentz force have two degrees of freedom—three dimensions minus one constraint ($\nabla \cdot \mathbf{B} = 0$)—and we also have two degrees of freedom in balancing the Lorentz force since we can adjust both the pressure and density fields independently of each other[5]. Note that in a fluid with a barotropic equation of state $\rho = \rho(P)$ we only have one degree of freedom in adjusting the pressure and density fields, so that depending on the context it may be either more difficult or impossible to construct an equilibrium. This question was explored in detail by Mitchell *et al* (2015) who concluded that the rudimentary considerations here seem to be correct.

In the special case without gravity we see from (8.42) that pressure is constant along field lines. Also, it appears at first glance that the gradients in thermal pressure $P$ and magnetic pressure $B^2/8\pi$ must be comparable. One way out of that is to have an approximately force-free field, i.e. a field where current and magnetic field are almost parallel so that the Lorentz force is very small. This would allow us to have a much higher magnetic pressure than thermal pressure. Also note that without gravity, density is no longer relevant for the structure of the equilibrium, meaning that we now only have one scalar field $P$ to balance the Lorentz force with its two degrees of freedom, just as in the case above with gravity and a barotropic equation of state.

---

[5] Strictly speaking, adjusting the density field will affect the gravitational field $\mathbf{g}$, but if only small adjustments to the non-magnetic equilibrium are needed it is hard to imagine that changes in $\mathbf{g}$ will prevent the existence of an equilibrium.

# Exercises

## 8.1 Field amplification

This problem examines how a magnetic field can be amplified in a given velocity field. The effect of the magnetic field on the velocity field is ignored, which is called the *kinematic regime*. Assume ideal MHD, i.e. perfect flux freezing.

(a) An initially uniform field $B_0\hat{\mathbf{y}}$ evolves in a shear flow where $\mathbf{u} = ay\hat{\mathbf{x}}$ where $a$ is a constant. Find an expression for the field at time $t$.

(b) Consider a magnetic field in a volume with some velocity field bounded by a stationary surface of fixed magnetic field, outside which the velocity field is zero. Initially, the field is in the lowest energy state possible, i.e. it is a potential field. Furthermore assume that the initial field is uniform in strength and direction, and ignore one of the dimensions perpendicular to the field. By relating the strength of the field to the distance between neighbouring field lines, argue that the lengthening of the field lines which results from 'stirring' of the fluid inevitably leads to higher energy.

(c) With the help of the continuity equation and some vector identities, write the induction equation in terms of $\mathrm{d}(\mathbf{B}/\rho)/\mathrm{d}t$. Comment on the physical meaning. (Hint: the same was done before for the vorticity equation.)

## 8.2 Tension of a flux tube

Imagine a flux tube of circular cross-section with radius $a$ and length $l$ containing a uniform magnetic field $B$.

(a) While holding $a$ constant, by considering the increase in magnetic energy while increasing the length by a *small* quantity, calculate the tension $T_{\mathrm{mag}}$ of the tube (in units of energy per unit length, i.e. force).

(b) The tube is in equilibrium in the lateral direction with its unmagnetised surroundings, giving rise to a difference in gas pressure inside and outside the tube of magnitude $B^2/8\pi$. Now calculate the total energy required to stretch the tube (again at constant $a$) by considering not only the increase in magnetic energy but also the $P\,\mathrm{d}V$ work done against the external gas in increasing the volume of the tube, minus that work done by the gas in the tube, showing that the tension is now double that found in part (a). Note that this is often neglected in the literature and that the tension calculated in part (a) is often used erroneously.

(c) Again assuming equilibrium in the lateral direction, calculate the tension in the tube again by considering the change in magnetic energy during a stretch at *constant volume*, so that the gas does no work, and constant magnetic flux.

(d) The tube is connected to itself in a circular loop of radius $r$ where we can assume $r \gg a$. The Lorentz force is now non-zero in the interior of the tube and points towards the centre of the circle. By integrating this curvature force, the second term in (8.21), over the entire volume of the tube, one finds the magnetic energy change during an infinitesimal change in $r$. Using the fact that $dl = 2\pi\, dr$, show that this gives the same tension in the tube as that found in parts (b) and (c), arguing that the Lorentz force at the surface of the tube can be neglected if the volume remains constant.

**8.3 MHD waves**

Starting from equations (8.27) to (8.31), derive the dispersion relations of waves where the wavevector is both parallel and perpendicular to the magnetic field, finding the phase and group velocities in each case and showing that the waves are non-dispersive (that is, speed does not depend on frequency). In addition, find the relation between the energy associated with the perturbation to the magnetic field $\delta\mathbf{B} \cdot \mathbf{B}/4\pi$, the perturbation to the gas, and the kinetic energy, showing that the magnetic and kinetic perturbations are equal in the parallel case, and that in the perpendicular case the kinetic energy accounts for one half of the energy and the other two forms account for the other half.

# References

Braithwaite J 2015 From pulsar scintillations to coronal heating: discontinuities in magneto-hydrodynamics *Mon. Not. R. Astron. Soc.* **450** 3201–10

Mitchell J P, Braithwaite J, Reisenegger A, Spruit H, Valdivia J A and Langer N 2015 Instability of magnetic equilibria in barotropic stars *Mon. Not. R. Astron. Soc.* **447** 1213–23

Van Doorsselaere T, Nakariakov V M and Verwichte E 2008 Detection of waves in the solar corona: kink or Alfven? *ApJ* **676** L73

Woltjer L 1958 A theorem on force-free magnetic fields *Proc. Nat. Acad. Sci.* **44** 489–91

# Chapter 9

## MHD: astrophysical contexts

We now illustrate the principles introduced in the last chapter and look at some processes and phenomena in the astrophysical contexts of the solar corona, jets and accretion discs.

### 9.1. The solar corona

On the photosphere of the Sun, we observe magnetic fields in the quiescent regions which are structured on the granulation scale (~1000 km) and hundreds of gauss in strength, and in addition we see active regions with sunspots of sizes 10 to 100 times the granulation scale in which the magnetic field is in the range 1–3 kG. The thermal energy density at the photosphere is about the same as a magnetic field of ~kG, which we call the 'equipartition field strength'. Since the flow speeds at and just below the photosphere are roughly sonic, the kinetic energy density is about the same. This explains why a field of at least around 1 kG is required to have much effect on the appearance of the photosphere. The thermal energy density increases rapidly below the photosphere and decreases rapidly above it—more rapidly than the magnetic energy density (pressure falls exponentially whereas the magnetic field tends to fall geometrically on large scales) so we can think of a $\beta = 1$ surface which lies at or just above the photosphere. Below this surface, the magnetic field has a rather subtle effect on the flow of gas; above this surface the magnetic field dominates. This region above the photosphere is called the corona (although strictly speaking the two are separated by the chromosphere and the transition region); see figure 9.1. The corona has a temperature of around 1–2 million K, which contrasts to the photospheric temperature of 5800 K—the origin of this high temperature, the 'coronal heating problem', is one of the best-known unsolved problems in astrophysics. It is generally agreed that the magnetic field transports energy through the photosphere and that it is converted from magnetic to thermal form in the corona; what is not understood is how the magnetic energy is dissipated. Theories normally invoke either reconnection or excitation and dissipation of magnetic waves

doi:10.1088/978-1-6817-4597-8ch9

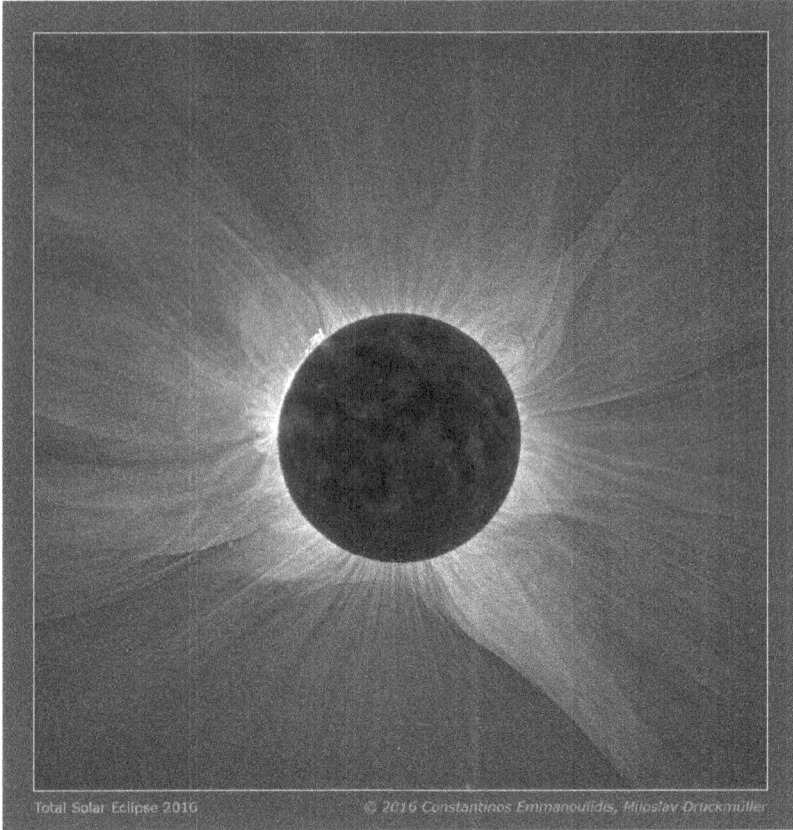

**Figure 9.1.** The solar corona photographed in Indonesia during the eclipse of 3 September 2016. (Reproduced with permission of M Druckmüller.)

(Tomczyk *et al* 2007; van Ballegooijen *et al* 2011; Hansteen *et al* 2015; De Moortel and Browning 2015). Table 9.1 lists some relevant parameters.

Looking at (8.34) it appears that since $\beta$ is low the Lorentz force should be much greater than the pressure gradient force. In addition, the gravitational force is comparable to the pressure force and acts only in the vertical direction. The Lorentz force must therefore be balanced by the inertia term on the left-hand side of the momentum equation, meaning that flow speeds are comparable to the Alfvén speed. However, structures such as those in figure 9.2 are observed to last for anything up to weeks, much greater than the Alfvén timescale $\tau_A = H_p/v_A \sim 30$ s. The only way out of this (as mentioned briefly in section 8.9) is for the Lorentz force to be reduced by having the current and magnetic field almost parallel to each other—we call this a 'force-free' field. The properties of these fields are explored in the next section. Often however we observe loop structures in the corona which, having apparently been in such a force-free equilibrium for some time, suddenly depart from equilibrium and convert much of their magnetic energy into heat on a timescale comparable to the Alfvén timescale. This process is explored in section 9.1.3.

**Table 9.1.** Very approximate parameter values in the solar corona.

| Parameter | Value |
|-----------|-------|
| T | $10^6$ K |
| $\rho$ | $10^{-15}$ g cm$^{-3}$ |
| P | $10^{-1}$ erg cm$^{-3}$ |
| g | $3 \times 10^4$ cm s$^{-2}$ |
| B | 10 G |
| $H_p$ | $3 \times 10^9$ cm |
| plasma-$\beta$ | $3 \times 10^{-2}$ |
| $c_s$ | $10^7$ cm s$^{-1}$ |
| $v_A$ | $10^8$ cm s$^{-1}$ |

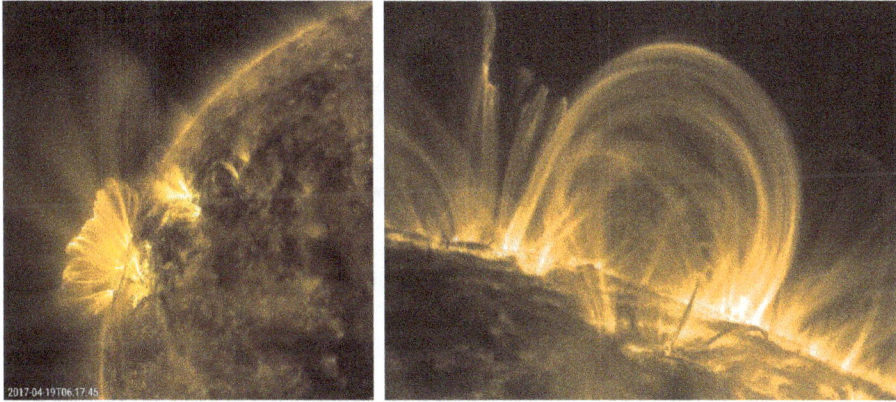

**Figure 9.2.** Images of coronal loops taken in the Fe IX line at 171 Å by TRACE. (Image credit: NASA/GSFC/ Solar Dynamics Observatory.)

### 9.1.1. Force-free and potential fields

In any low $\beta$ plasma we see from (8.34) that the Lorentz force cannot apparently be balanced by the pressure gradient and it is difficult to imagine a gravitational field with the necessary geometry; balancing the Lorentz force with inertia, the term on the left-hand side of the momentum equation (8.32), would mean Alfvénic flow speeds and nothing approaching an equilibrium. We therefore speak of a 'force-free' field, where the current and magnetic field are almost parallel so that $|(\nabla \times \mathbf{B}) \times \mathbf{B}| \ll B^2/L$. This gives:

$$\nabla \times \mathbf{B} = \alpha \mathbf{B}, \tag{9.1}$$

$$\mathbf{B} \cdot \nabla \alpha = 0, \tag{9.2}$$

where the second equality comes from taking the divergence of the first and using the solenoidal condition $\nabla \cdot \mathbf{B} = 0$; it means that $\alpha$ is constant along field lines. In other words, as we follow a field line we see that the neighbouring lines curve around it in the same sense all the way along the line—force-free fields are 'twisted' in some sense. If $\alpha$ is a constant everywhere, we can write the Helmholtz equation $(\alpha^2 + \nabla^2)\mathbf{B} = 0$ by taking the curl of (9.1).

A special case is where $\alpha = 0$, which we call a *potential* or *curl-free* field, where the current vanishes. This is obviously the case in a vacuum, and is a good approximation in some other astrophysical contexts. We call it a potential field because being curl-free we can express it as the gradient of a scalar potential $\mathbf{B} = \nabla\phi$. Since the divergence of the field is zero, we have the Laplace equation $\nabla^2\phi = 0$. This we can solve if we know the normal component $\mathbf{n} \cdot \mathbf{B} = \mathbf{n} \cdot \nabla\phi$ everywhere on the boundary of the domain.

There is a theorem which states that no equilibrium can be force-free everywhere, called the *vanishing force-free field theorem*. Imagine a force-free equilibrium in a region of volume $V$ surrounded by an unmagnetised region, and imagine an isotropic expansion or contraction of the region. Under such a change, the position vector of any fluid element changes from $\mathbf{r}$ to $\mathbf{r}'$ and the field from $\mathbf{B}$ to $\mathbf{B}'$. In a uniform expansion by a factor $a$ we have $\mathbf{r}' = a\mathbf{r}$ and from flux conservation ($\phi = Br^2$ is constant) we see that $r'^2\mathbf{B}'(\mathbf{r}') = r^2\mathbf{B}(\mathbf{r})$. The energy of the field after the expansion is

$$E' = \int_{V'} \frac{B'^2}{8\pi}\mathrm{d}V' = \frac{1}{a} \int_V \frac{B^2}{8\pi}\mathrm{d}V \tag{9.3}$$

since $\mathrm{d}V' = a^3\,\mathrm{d}V$ and $\mathbf{B}' = a^{-2}\mathbf{B}$. The region will therefore expand until either some force opposes it, at which point it is no longer force-free, or it reaches infinite extent and $E \to 0$. (More generally, anything with positive energy will tend to expand.) A force-free region must be subject to forces on its boundary.


### 9.1.2. Energy minima

Imagine a fixed volume with a given normal field component at the boundary; outside the volume there is no motion and no change in the magnetic field, and no energy is injected into the volume in mechanical or other form. One can solve for a potential field in the volume and the solution is unique. Furthermore, it can be shown that this field has the lowest energy of all which satisfy the boundary conditions. This can be seen by the following argument. Magnetic energy can be converted into other forms in two ways: into kinetic energy via the Lorentz force (this can work in both ways) and into heat energy via Joule heating (one-way conversion, see section 8.6). Any magnetic field with non-zero current will continuously be losing energy into heat, and this cannot be replaced by conversion from kinetic, since we know from the second law of thermodynamics that the heat energy cannot entirely be converted back to kinetic. There can be some back-and-forth flow of energy between magnetic and kinetic, i.e. oscillations, but these necessarily involve the magnetic field being always (except perhaps fleetingly) in a non-potential state with non-zero Joule heating. The energy of any non-potential field must therefore drop until the Joule heating vanishes entirely, i.e. when the

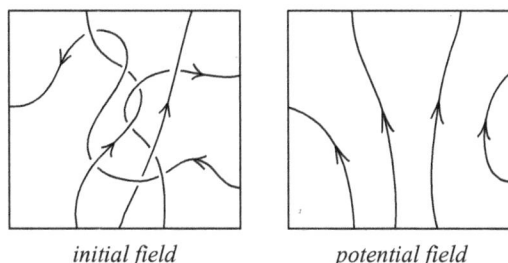

*initial field*          *potential field*

**Figure 9.3.** In a volume of fluid with finite conductivity and with fixed boundaries, an initial field of arbitrary complexity and topology eventually relaxes to a potential field.

current is zero everywhere—*quod erat demonstrandum*. If there is no means to convert kinetic energy directly into heat (viscosity $\nu = 0$) then all that can remain are acoustic oscillations propagating parallel to the field lines (a special and unlikely situation), otherwise all kinetic and all *free* magnetic energy is ultimately converted into heat. In summary, any field in a medium of finite conductivity in a volume with fixed boundaries will relax to this potential field. Once this state is reached, there is no way to extract the remaining $B^2/8\pi$ energy without changing the boundaries; see figure 9.3.

The situation is different if flux-freezing holds perfectly, i.e. the medium has infinite conductivity. If we also assume zero kinetic diffusivity there is no conversion of either magnetic or kinetic energy into heat and oscillations will not be damped. If viscosity is present, however, kinetic energy will be converted to heat and any oscillations or 'sloshing' will therefore result in energy loss from magnetic/kinetic into heat. This can only result in a stationary final state, which is presumably an energy minimum, i.e. minimum magnetic energy and zero kinetic energy. Perturbing this state with a displacement field $\boldsymbol{\xi}$ should therefore not affect the magnetic energy, since the energy should rise fastest quadratically away from the minimum. The only way this can be possible is if the energy minimum is force-free. Force-free states are no longer unique in the same way as the potential states—here, unlike the finite conductivity case, flux freezing holds and so the final state now depends not only on the normal component of the field at the boundary but on the topology of the field, i.e. how the field lines entering the volume connect to those leaving[1].

Astrophysically, both of these cases occur in many situations. An example of a potential field would be the volume outside an intermediate mass main-sequence star—inside the star the conductivity is high and the Ohmic timescale is very long, so the star can contain a long-lived equilibrium which gives an essentially fixed normal component at the surface. Outside the star the conductivity and therefore the Ohmic timescale are much lower than in the interior so that the field relaxes to a potential field. In the *solar* corona, on the other hand, the temperature and therefore

---

[1] In fact even given both the boundary and topological constraints, the uniqueness of the force-free field is not obvious since local energy minima seem plausible, but that is outside the scope of this course.

conductivity are higher than in the intermediate mass star. Moreover, the field at the surface is not static but moves on timescales of minutes to weeks, much shorter than the Ohmic timescale, so that the field does not have time to relax to potential. In addition, since the mean free path is so large, the kinetic diffusivity is large—very much larger than the magnetic diffusivity, so that kinetic energy can be removed but flux freezing holds. The field in the solar corona is indeed observed to be very close to force-free.

### 9.1.3. Reconnection

We saw above in section 8.9 and from equation (8.33) that the timescale over which the magnetic diffusivity acts is $\tau_{\mathrm{diff}} \sim L^2/\eta$. However, we see in many astrophysical contexts such as the solar corona that changes in global magnetic topology, i.e. deviations from flux-freezing, and the associated dissipation of magnetic energy can occur on much shorter timescales. For instance, energy is released during solar flares over timescales of seconds and minutes although $\tau_{\mathrm{diff}} \gtrsim 10^6$ yr, assuming a standard Spitzer conductivity. There are two possible reasons for this. The first is that somehow the diffusion is locally brought to work on shorter length scales than the global length scale $L$, the second is that there is some 'anomalous resistivity'—higher than the standard resistivity—perhaps when the current density is particularly high; plasma instabilities may also be involved. It seems likely that in some situations both mechanisms must be invoked.

Producing structure on small length scales from a global configuration initially lacking such small scales is a common phenomenon in physics. For instance, to mix two paints together so that they combine on the microscopic scale, it is sufficient to stir with a large spoon. In a turbulent flow, large-scale driving leads to the appearance of structure on a scale sufficiently small that the viscous dissipation timescale is equal to the flow timescale[2]. Similarly, in a shock a small length scale is produced to allow fast diffusive conversion of kinetic to heat energy. In the example of solar flares, we do not see turbulence; rather, the field is thought to dissipate in thin current sheets or otherwise localised features which separate regions with different magnetic field. This process is known as 'reconnection'.

The first model of reconnection, the *Sweet–Parker* mechanism, is illustrated in figure 9.4. Material flows perpendicular to the field lines at speed $v_0$ towards the current sheet. This speed is equated to the 'diffusion speed' within the current sheet, which from the induction equation is equal to $\eta/\delta \approx v_0$. Assuming incompressibility, conservation of mass gives $v_0 L \approx v_* \delta$. Now, if we consider the force balance in the $y$ direction it is clear that there is an excess thermal pressure at the centre of the current sheet, where the magnetic field vanishes, equal to the magnetic pressure at the boundaries of the sheet, i.e. $B^2/8\pi$. The material is accelerated in the $x$ direction by this thermal pressure and escapes at the ends of the sheet; we have from the force

---

[2] Opinions differ as to how this works.

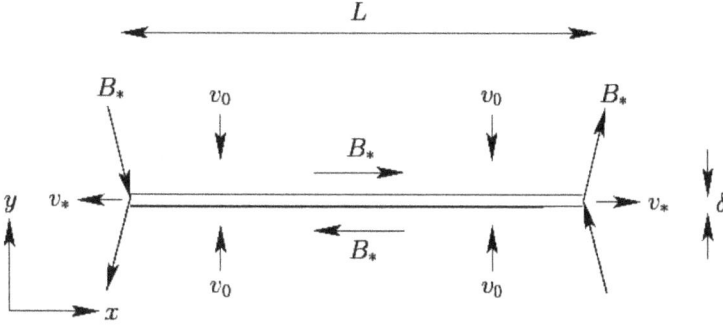

**Figure 9.4.** The Sweet–Parker reconnection mechanism. Regions of opposing magnetic field $B_*$ are brought together, separated by a thin sheet of thickness $\delta$.

balance that $\rho v_*^2/2 \approx B^2/8\pi$ which can be rearranged to $v_* \approx v_A = B/\sqrt{4\pi\rho}$. Now we can solve for the reconnection velocity $v_0$:

$$v_0 \approx \frac{\eta}{\delta} \approx \frac{v_A \eta}{v_0 L} \qquad \Longrightarrow \qquad \frac{v_0}{v_A} \approx \sqrt{\frac{\eta}{v_A L}} = \mathrm{Re_A}^{-1/2}, \tag{9.4}$$

where $\mathrm{Re_A}$ is the Alfvénic Reynolds number. This model produces reconnection speed ratios of $v_0/v_A \sim 10^{-6}$ in the solar corona and other astrophysical plasmas, which is unfortunately rather less than the generally observed value of $\sim 0.1$. One way out of this is the Petschek reconnection model in which most of the energy is dissipated in standing shocks attached to a small central Sweet–Parker-like diffusion region. Some kind of anomalous resistivity probably also plays a role, as well as three-dimensional effects. Finally, note that reconnection does not only convert magnetic energy into kinetic and thermal, but it also accelerates particles up to relativistic speeds—this is observed in the solar corona. In general, magnetic fields are ingredients in most cosmic-ray acceleration mechanisms. For more on reconnection the reader is recommended to look at Uzdensky (2014) or Cassak *et al* (2017) and references therein.

## 9.2. Jets: launching, collimation and instabilities

Jets are found in many astrophysical accretion settings, for example protostars, neutron stars and active galactic nuclei. In this section we examine the magneto-centrifugal model of jet launching (Blandford and Payne 1982) and collimation.

### 9.2.1. Launching

Imagine a Keplerian disc around a central object threaded by a magnetic field (figure 9.5). The field which emerges from the disc is 'ordered' in some sense. It is natural to assume that the field strength in the disc is greater towards the centre, and for this reason the field should be angled outwards above and below the disc so that the contrast in field strength outside the disc is reduced, which represents the lowest

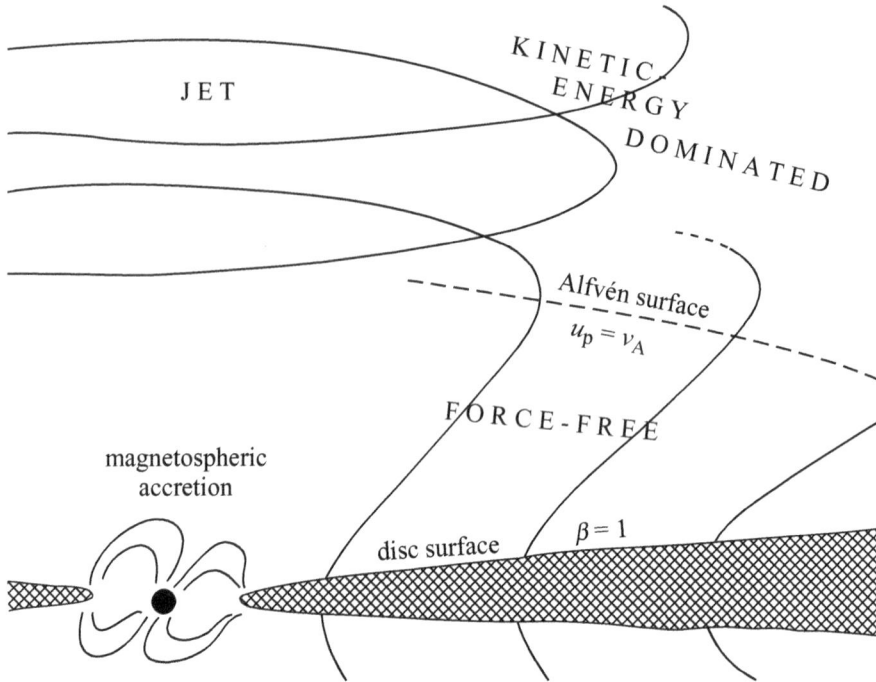

**Figure 9.5.** A disc wind and jet. Inside the disc $\beta \gg 1$ and the thermal energy dominates over the magnetic field. Gas evaporates from the disc and is accelerated centrifugally along the field lines; in this region the field is force-free. At the Alfvén surface the gas reaches the Alfvén speed. Beyond this surface the poloidal kinetic energy dominates and the magnetic field is wound up into a spiral. Meanwhile gas accretes onto the star from the inner edge of the disc via some complicated process in the star's magnetosphere.

magnetic energy state (recall that energy density is proportional to the square of field strength).

We now allow material to evaporate from the disc. Just above the disc this material will have a low density in the sense that the magnetic energy density $B^2/8\pi$ is much greater than the thermal, i.e. $\beta \ll 1$, and also than the kinetic, i.e. $B^2/8\pi \gg \rho u_p^2/2$. (The poloidal component of the velocity $u_p$ reaches the Alfvén speed; the toroidal (azimuthal) part of the velocity $u_t$ was already supersonic and super-Alfvénic from the orbital motion.) This means that the field must be force-free and that it is little affected by the material. Flux freezing requires that the material flows along the field lines, and since the field lines point away from the central object, the material is centrifugally accelerated away from the disc. It can be shown that if the angle between the field and the vertical exceeds some threshold then this centrifugal acceleration exceeds the downwards gravitational acceleration. The material is forced to co-rotate with the magnetic field.

As the material is accelerated its kinetic energy density eventually exceeds the magnetic, i.e. $\rho u_p^2/2 > B^2/8\pi$ or alternatively $u_p > v_A$. We call the location of this transition the 'Alfvén surface', which is analogous to the sonic point in non-magnetised flows such as in the nozzle of a rocket engine (see section 3.3). Flux freezing holds on both sides, but whereas inside the Alfvén surface the flow follows the field lines, outside the Alfvén surface the field lines follow the flow. This means that the material ceases to co-rotate with the foot points of the field lines; rather, the field lines are 'wound up' so that a significant toroidal (azimuthal) component $B_t$ is produced.

### 9.2.2. Collimation

There is much evidence that jets are collimated, sometimes with a very small opening angle. An example is shown in figure 9.6. Exactly how the collimation works is still a matter of debate.

One possibility is that it is the poloidal component of the magnetic field that collimates the flow. If, for simplicity, we assume that the field component emerging normal to the disc is of uniform sign and its strength varies with cylindrical radius as $B_z = B_0(\varpi^2/\varpi_0^2 + 1)^{-1/2}$, and then assume that the field above the disc is curl-free (i.e. zero-current, force-free with $\alpha = 0$) we have the field illustrated in figure 9.7. Near the disc the lines are inclined away from the centre because of the greater field strength at the centre, but further from the disc they tend towards the vertical because of the fact that flux per unit cylindrical radius increases outwards—in other words, most of the flux is threaded through the outer disc and so from a distance the

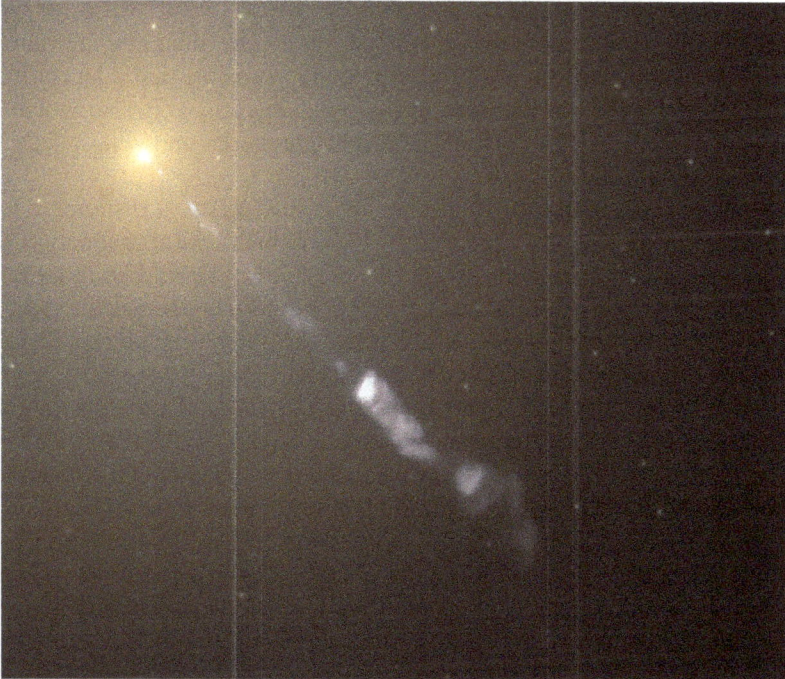

**Figure 9.6.** The jet coming from the nucleus of M87, the largest of the large elliptical galaxies in the Virgo cluster. How such a fine collimation is achieved is a matter of debate. (Image credit: J A Biretta *et al* Hubble Heritage Team (STScI/AURA), NASA.)

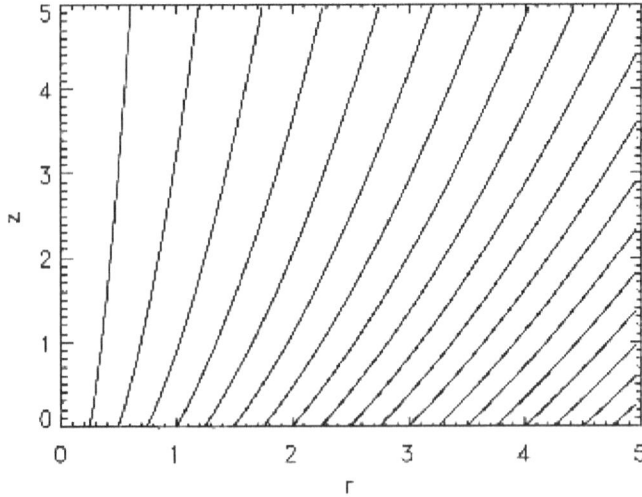

**Figure 9.7.** The field above a disc from which a vertical field component $B_z = B_0(\varpi^2/\varpi_0^2 + 1)^{-1/2}$ emerges, assuming the field is curl-free. Note that the field lines curve towards the vertical. (Reproduced with permission from Spruit *et al* 1997).

inner part is of lesser importance. We expect that $B_z \propto \Sigma$ where $\Sigma$ is the column density of the disc, so this picture of flux increasing outwards is realistic as long as $\varpi\Sigma$ increases outwards, i.e. $\partial \ln \Sigma / \partial \ln \varpi > -1$.

In this picture of collimation, the jet emerging from the inner disc is collimated by the magnetic field from the outer disc. It also seems plausible that a jet is forced into a narrow passage by the gas pressure in the outer regions. It is also worth though looking at the evolution of the magnetic field inside the jet and what effect it might have on the collimation.

Imagine a jet with circular cross-section of radius $a(z)$ at a distance $z$ from the central object (outside the Alfvén surface). It contains a spiral magnetic field with toroidal and poloidal components $B_t$ and $B_p$. If the toroidal and poloidal fluxes are both conserved as the material moves away from the source, then as the radius of the jet changes we have $B_t \propto a^{-1}$ and $B_p \propto a^{-2}$. After the initial acceleration phase the jet will expand ballistically with $a \propto z$ in the absence of significant pressure or magnetic forces. Obviously in the non-magnetic case, if the thermal pressure in the jet is greater than that in the surroundings, the jet will expand faster than $a \propto z$—the jet will be 'flared'. The poloidal component of the field in the jet will also tend to make the jet flare as it exerts a pressure $B_p^2/8\pi$ on the jet's surroundings; the toroidal field exerts no pressure because its pressure and tension forces are equal and opposite. Another way of thinking about this is as follows. As the jet expands the toroidal and poloidal fluxes per unit length $\Phi_t = aB_t$ and $\Phi_p = \pi a^2 B_p$ are conserved. The energies per unit length of the jet in the toroidal and poloidal components of the field are:

$$E_t = \pi a^2 \frac{B_t^2}{8\pi} = \frac{\Phi_t^2}{8}, \tag{9.5}$$

$$E_p = \pi a^2 \frac{B_p^2}{8\pi} = \frac{\Phi_p^2}{8\pi^2 a^2}. \tag{9.6}$$

While $E_t$ is indifferent to a change in $a$, $E_p$ drops if the jet expands and so the poloidal field will obviously tend to drive an expansion of the jet. It is therefore necessary to have some external pressure for collimation to occur—one can imagine that with some constant pressure in the ambient medium the jet would be flared near the source and then further away would settle at constant $a$ when its internal pressure $P_{jet} + B_p^2/8\pi$ is equal to the external pressure.

Since it is discussed a lot in the literature, it is worth looking in slightly more detail at the collimating effect of the toroidal field. A jet with $B_t = B_t(\varpi)$ ($\varpi$ is the cylindrical radius coordinate) carries a current $\mathbf{J} = \hat{z}(c/4\pi\varpi)\partial(\varpi B_t)/\partial\varpi$ and so the Lorentz force is

$$\mathbf{F}_{Lor} = -\hat{\varpi}\frac{B_t}{4\pi\varpi}\frac{\partial(\varpi B_t)}{\partial\varpi}, \tag{9.7}$$

where $\hat{\varpi}$ is the unit vector in the $\varpi$ direction. At least near the axis this force must be directed inwards: this effect is often referred to as 'hoop stress'. This has led to the misleading concept of 'self-collimation', according to which a jet can be collimated by its own toroidal magnetic field. The problem with this is that to avoid having the energy diverge towards infinity, the partial derivative must change sign at some radius outside which the Lorentz force is directed outwards, requiring in effect some external pressure support. In fact it should be obvious anyway that something with positive energy isn't going to confine itself. Self-collimation is a fully paid-up member in the club of extraordinary popular delusions.

### 9.2.3. Instability

We saw above that as a jet expands the poloidal field falls faster than the toroidal, which leads eventually to an instability driven by the free energy in the toroidal field. It can be shown that the dominant modes have azimuthal wavenumbers $m = 0$ (sausage mode) and $m = 1$ (kink mode); here we look at a simple derivation of the instability criterion for the $m = 0$ mode.

A purely toroidal field is some function of cylindrical radius, $B = B(\varpi)$. Imagine two thin annuli at radii $\varpi$ and $\varpi + \delta\varpi$ with magnetic fields $B$ and $B + \delta B$, each of area $A$ and therefore of thicknesses $A/(2\pi\varpi)$ and $A/(2\pi(\varpi + \delta\varpi))$. The energy per unit length jet of the magnetic field in the annuli is

$$E = \frac{A}{8\pi}[B^2 + (B + \delta B)^2]. \tag{9.8}$$

We now exchange adiabatically the positions of the two annuli, keeping the volume of each constant. In general, the most unstable modes of any instability will be incompressible (density unchanged), since compressing the gas will require work to be done by the magnetic field. Since the volumes of the annuli remain the same, the total thermal energy is unchanged. Since flux is conserved, the new fields at locations $\varpi$ and $\varpi + \delta\varpi$ are $(B + \delta B)\varpi/(\varpi + \delta\varpi)$ and $B(\varpi + \delta\varpi)/\varpi$, so that the new energy is

$$E + \delta E = \frac{A}{8\pi}\left[\left(\frac{(B + \delta B)\varpi}{\varpi + \delta\varpi}\right)^2 + \left(\frac{B(\varpi + \delta\varpi)}{\varpi}\right)^2\right]. \tag{9.9}$$

For stability we need the exchange to have increased the energy, i.e. $\delta E > 0$. Subtracting (9.8) from (9.9) and dividing by $A/8\pi$ we have

$$\left(\frac{(B + \delta B)\varpi}{\varpi + \delta\varpi}\right)^2 + \left(\frac{B(\varpi + \delta\varpi)}{\varpi}\right)^2 - B^2 - (B + \delta B)^2 > 0, \tag{9.10}$$

$$\left(1 + 2\frac{\delta B}{B} + \frac{\delta B^2}{B^2}\right) + \left(1 + 2\frac{\delta\varpi}{\varpi} + \frac{\delta\varpi^2}{\varpi^2}\right)\left[\left(1 + 2\frac{\delta\varpi}{\varpi} + \frac{\delta\varpi^2}{\varpi^2} - 1\right)\right.$$
$$\left. - \left(1 + 2\frac{\delta B}{B} + \frac{\delta B^2}{B^2}\right)\right] > 0, \tag{9.11}$$

where the zeroth and first-order terms cancel; keeping only the second-order terms we have

$$\frac{\delta\varpi}{\varpi} - \frac{\delta B}{B} > 0 \qquad \text{or} \qquad \frac{\partial \ln B}{\partial \ln \varpi} < 1. \tag{9.12}$$

This corresponds to the result of Tayler (1957). For obvious reasons, this is known as an 'interchange' mode. A more general treatment including the non-axisymmetric modes reveals that $m \geqslant 1$ modes are stable if $\partial \ln B/\partial \ln \varpi < m^2/2 - 1$, meaning that the $m = 1$ mode (the 'kink mode') sets in first. Note that to avoid a current singularity we need $\partial \ln B/\partial \ln \varpi \geqslant 1$ on the axis, so it is impossible in practice to construct a toroidal field which is stable everywhere. The growth timescale of all modes is comparable to the dynamical timescale, i.e. the Alfvén timescale $\varpi/v_A$.

A jet contains not only a toroidal component, of course, but also an axial component $B_z$ which can help to stabilise the jet against this instability. We can see approximately how strong this component needs to be by means of the following energy argument. It is clear that as the instability grows, work needs to be done against the axial component of the field and for stability this must be greater than the energy released from the toroidal part of the field via the instability. Now, since $\ddot\xi = \sigma^2\xi$ where the growth rate $\sigma = v_A/\varpi$, the $m = 1$ mode releases an energy per unit volume equal to $\frac{1}{2}\rho\ddot\xi\xi = \frac{1}{2}\rho\sigma^2\xi^2$. As the axial field lines are stretched, they exert a restoring force $\frac{1}{4\pi}B_z^2\xi/l_z^2$ where $l_z$ is the length scale of the instability in the $z$ direction, which is equal to $1/k_z = \lambda_z/2\pi$. The work to be done therefore is $\frac{1}{8\pi}B_z^2\xi^2/l_z^2$, which for stability needs to be greater than the energy released so that

$$\frac{1}{8\pi}B_z^2\xi^2\bigg/l_z^2 > \frac{1}{2}\rho\sigma^2\xi^2, \tag{9.13}$$

which can be rewritten as

$$\frac{B_z^2}{4\pi\rho}\frac{1}{l_z^2} > \sigma^2, \tag{9.14}$$

$$\frac{B_z^2}{4\pi\rho}\frac{\varpi^2}{l_z^2} > \frac{B_\varpi^2}{4\pi\rho}, \tag{9.15}$$

$$\frac{B_z}{B_\varpi} > \frac{\lambda_z}{2\pi\varpi}. \tag{9.16}$$

The axial field therefore stabilises the shortest wavelengths, because it is the shortest wavelengths which have to bend the axial field lines to a greater degree for a given $|\xi|$. Another way of expressing this result is that instability sets in for wavelengths greater than the distance over which a field line makes one full circle around the jet. This is often referred to in the literature as the *Kruskal–Shafranov* condition. The ratio $\lambda_z/\varpi$ could be very large in a narrow jet so this instability is expected to be present in essentially all collimated jets; work continues therefore on its non-linear development.

## 9.3. Angular momentum transport in discs

Accretion discs are found in many astrophysical contexts, such as during star formation, mass transfer in binary systems and accretion of gas onto supermassive black holes. When matter is accreted, it falls deeper into a gravitational potential well and energy must be lost from the system, generally via radiation. However, it is not possible for a disc of material in isolation to accrete entirely onto the central object because of angular momentum conservation—angular momentum must be removed from the accreting material by transfer to other material which does not accrete. Therefore the lowest-energy end-state of a disc in isolation is to have an infinitesimally small amount of mass move outwards towards infinity and infinite specific angular momentum and the rest of the mass accreted onto the central object. This requires transport of angular momentum outwards[3]. Since all systems like to relax to energy minima, we should expect to find some mechanism operating in discs for the outward transport of angular momentum. According to the model of Shakura and Sunyaev (1973), the observational properties of accretion discs can be reproduced well by imagining there is some viscous stress equal in magnitude to some fraction $\alpha$ of the thermal pressure $P$. However, microscopic viscosity is far too small to account for values $\alpha \sim 0.1$ inferred from the observations; we turn our attention therefore to possible instabilities which could generate turbulence and the resultant 'turbulent viscosity'.

---

[3] Transport is outwards in the Langrangian sense of each fluid element transfering its angular momentum to its outside neighbour. Imagining the rate of change of angular momentum inside a fixed volume containing the central object (whose angular momentum is increasing as it is spun up and becomes more massive) and part of a steady-state disc, we see that net transport must be *inwards*; in other words, the inwards advection of angular momentum exceeds the transport though turbulent stress, albeit only by some small amount.

An example of a shear instability in a differentially rotating flow is the Rayleigh instability, an interchange instability. Imagine exchanging two annuli of equal volume and density between two radii $\varpi$ and $\varpi + \delta\varpi$ which are initially moving with angular velocities $\Omega$ and $\Omega + \delta\Omega$. The kinetic energies before and after the exchange are $E$ and $E + \delta E$:

$$E = \frac{\rho V}{2}[\varpi^2\Omega^2 + (\varpi + \delta\varpi)^2(\Omega + \delta\Omega)^2], \tag{9.17}$$

$$E + \delta E = \frac{\rho V}{2}\left[\varpi^2(\Omega + \delta\Omega)^2\left(\frac{\varpi + \delta\varpi}{\varpi}\right)^4 + (\varpi + \delta\varpi)^2\Omega^2\left(\frac{\varpi}{\varpi + \delta\varpi}\right)^4\right], \tag{9.18}$$

$$\delta E \approx 2\rho V\varpi^2\Omega^2\delta\ln\varpi^2\left[2 + \frac{\delta\ln\Omega}{\delta\ln\varpi}\right] \tag{9.19}$$

so that the stability condition is $q \equiv \partial\ln\Omega/\partial\ln\varpi > -2$, or in other words that the specific angular momentum $\varpi^2\Omega$ increases outwards. An accretion disc ($q = -3/2$) is therefore stable to this mechanism. However, another kind of shear instability, the *magneto-rotational instability* (MRI) has the stability condition that $q > 0$. Although this instability has been known about for a long time (Chandrasekhar 1961), its relevance in accretion discs did not become clear until much more recently (Balbus and Hawley 1991; see also Hawley *et al* 2015).

### 9.3.1. MRI: physical mechanism and stability condition

The physical mechanism can be thought of in the following way, illustrated in figure 9.8. Consider two fluid elements, initially at the same radius, one above the

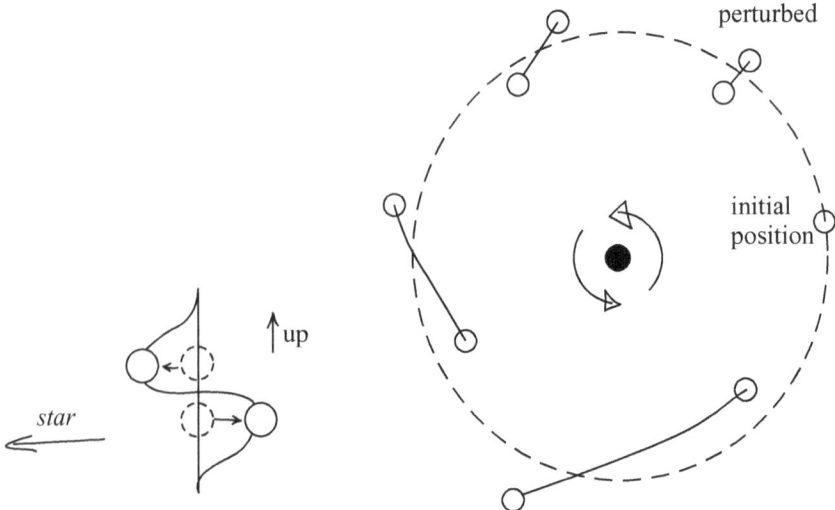

**Figure 9.8.** Left: the initial and perturbed locations of two fluid elements connected by a magnetic field line; the line between them is initially perpendicular to the plane of the disc. Right: viewed from above the disc, the evolution of the positions of the two fluid elements once they have been perturbed.

other so that they are threaded by the same field line. They are given a perturbation in the radial direction, in opposite senses. Initially their angular momenta remain unchanged so that the element which has been perturbed inwards moves faster than the other. As it moves forwards with respect to the other, the field line connecting them is stretched so that the inner element pulls on the outer, transferring angular momentum to it. This causes the inner element to move further inwards and the outer element to move further outwards. In addition, in some sense the field line can be thought of as a spring which oscillates at a frequency $v_A/\lambda$ where $\lambda$ is the wavelength. Clearly the spring has to have a lower intrinsic frequency than that at which it is driven in order that it can be stretched instead of oscillating, meaning that $v_A/\lambda < \Omega/2\pi$. The field has therefore to be 'weak' in some sense for the instability to proceed.

Another way of imagining the instability is the following. Consider a fluid element at radius $\varpi$ threaded by a vertical magnetic field $B$. In the rotating frame, in equilibrium the centrifugal force per unit mass at any radius $\Omega^2\varpi$ is balanced by some net inward-pointing force which in a disc is a combination of a pressure gradient and a gravitational force. The fluid element is now displaced a distance $\delta\varpi$ to a radius $\varpi + \delta\varpi$, while the magnetic field couples it to the other fluid elements on the same field line so that it retains its original angular velocity (angular momentum is transferred to it). The inward-pointing force at this new position is $(\Omega + \delta\Omega)^2(\varpi + \delta\varpi)$ but the centrifugal force is now $\Omega^2(\varpi + \delta\varpi)$. In addition there is a magnetic restoring force, giving the total (radial) force

$$
\begin{aligned}
F &= \Omega^2(\varpi + \delta\varpi) - (\Omega + \delta\Omega)^2(\varpi + \delta\varpi) - \frac{B^2}{4\pi\rho}\frac{\delta\varpi}{(\lambda_z/2\pi)^2} \\
&= \Omega^2\varpi\left[(1 + \delta\ln\varpi) - (1 + \delta\ln\varpi)^2(1 + \delta\ln\varpi) - \frac{v_A^2}{\Omega^2}\frac{\delta\ln\varpi}{(\lambda_z/2\pi)^2}\right] \\
&= \Omega^2\varpi\left[-2\delta\ln\Omega - \frac{v_A^2}{\Omega^2}\frac{\delta\ln\varpi}{(\lambda_z/2\pi)^2}\right] \\
&= -\Omega^2\delta\varpi(2q + \kappa_z^2),
\end{aligned}
\tag{9.20}
$$

where $\lambda_z$ is the wavelength of the perturbation in the vertical direction, $v_A^2 = B^2/4\pi\rho$ is the Alfvén speed and $\kappa_z = k_z v_A/\Omega$ is a dimensionless wavenumber where $k_z = 2\pi/\lambda$ is the vertical wavenumber. For stability we need $F < 0$ and therefore

$$
2q + \kappa_z^2 > 0, \tag{9.21}
$$

remembering that $v_A^2 = B^2/4\pi\rho$. For stability at all vertical wavenumbers, we clearly need $q > 0$, and in the unstable case the maximum unstable wavelength is given by $\kappa_z^2 = 2|q|$; as we shall see below, these estimates agree with the more rigorous treatment. Note that $F$ is equal to the acceleration $\partial^2(\delta\varpi)/\partial t^2$; we can replace $\partial^2/\partial t^2$ with $-\omega^2$ where $\omega$ is the oscillation frequency (if real) or the growth rate (if imaginary). This gives, from (9.20),

$$
-\omega^2 = -\Omega^2(2q + \kappa_z^2), \tag{9.22}
$$

so that the growth rate will generally be comparable to the rotation frequency. This not surprising, because the free energy source is the rotation. In this sense the instability is fundamentally different from the instability of toroidal fields described in section 9.3, whose free energy source is the magnetic field and whose growth time is comparable to the Alfvén timescale.

### 9.3.2. The dispersion relation

In our rotating fluid, locally we can look at a small corotating Cartesian volume at distance $\varpi_0$ from the centre, rotating at $\Omega_0$, and change variables to $x = \varpi - \varpi_0$; the azimuthal direction is $y$ and the vertical direction $z$. Since $\Omega(x) \approx \Omega_0 + qx\Omega_0/\varpi_0$, the bulk velocity in the rotating frame is $U \approx q\Omega_0 x$ (where $q = -3/2$ in the Keplerian case). In a rotating frame of reference, we must generally add to the momentum equation (8.32) both a Coriolis force $-2\rho\boldsymbol{\Omega}_0 \times \mathbf{v}$ (where $\mathbf{v}$ is the total velocity field) and a centrifugal force $\hat{\boldsymbol{\varpi}}\rho\varpi\Omega_0^2$, but here we can drop the centrifugal term, the steady part of the Coriolis force coming from the basic flow $U_y$ and the $x$-component of the gravitational term because they cancel each other, leaving just the Coriolis force associated with any additional velocity field $\mathbf{u} = \mathbf{v} - U\hat{\mathbf{y}}$ on top of the basic flow, $-2\rho\boldsymbol{\Omega} \times \mathbf{u}$. In an accretion disc we also have the vertical part of the gravitational acceleration $-\rho GMz\varpi_0^{-3}\hat{\mathbf{z}} = -\rho z\Omega_0^2\hat{\mathbf{z}}$, where the assumption is made that $z \ll \varpi_0$. This vertical stratification may be important in realistic discs, but we shall ignore it for the time being.

A general linear analysis of the MRI is quite involved, so we make two further simplifications: the initial magnetic field is of uniform strength and in the $z$ direction so that $\mathbf{B} = B\hat{\mathbf{z}}$, and we consider only axisymmetric modes, meaning that $\partial/\partial y = 0$. The most unstable modes, at least for a weak field, will be incompressible and we assume that in the following. The perturbation to the magnetic field is $B\mathbf{b}$, so that $\mathbf{b}$ is dimensionless. We now linearise the MHD equations by subtracting equilibrium (zero-order) terms in the momentum equation, and keeping terms to first-order in the perturbed quantities, ignoring the diffusive terms Beginning with the momentum equation we have

$$\partial_t\mathbf{u} + \mathbf{U} \cdot \nabla\mathbf{u} + \mathbf{u} \cdot \nabla\mathbf{U} = -\frac{1}{\rho}\nabla\delta P + \frac{1}{4\pi\rho}[(\nabla \times (B\mathbf{b})) \times \mathbf{B}$$

$$+ (\nabla \times \mathbf{B}) \times (B\mathbf{b})] - 2\boldsymbol{\Omega} \times \mathbf{u},$$

$$\partial_t\mathbf{u} + \hat{\mathbf{y}}u_x\partial_x U = -\frac{1}{\rho}\nabla\delta P + v_A^2(\nabla \times \mathbf{b}) \times \hat{\mathbf{z}} - 2\Omega\hat{\mathbf{z}} \times \mathbf{u}, \qquad (9.23)$$

$$\partial_t\mathbf{u} = -\frac{1}{\rho}\nabla\delta P + v_A^2(\nabla \times \mathbf{b}) \times \hat{\mathbf{z}}$$

$$- \Omega(2\hat{\mathbf{z}} \times \mathbf{u} + q\hat{\mathbf{y}}u_x),$$

noting that $\mathbf{U} \cdot \nabla\mathbf{u} = 0$ and $\nabla \times \mathbf{B} = 0$. We can now consider perturbations which vary in space and time as $e^{i(\mathbf{k}\cdot\mathbf{r}-\omega t)}$ so that $\partial_x = ik_x$ and so on:

$$-i\omega\mathbf{u} = -\frac{i\mathbf{k}}{\rho}\delta P + iv_A^2(\mathbf{k} \times \mathbf{b}) \times \hat{\mathbf{z}} - \Omega(2\hat{\mathbf{z}} \times \mathbf{u} + q\hat{\mathbf{y}}u_x), \qquad (9.24)$$

$$-i\psi\mathbf{u} = -i\kappa\frac{\delta P}{\rho v_A} + iv_A(\boldsymbol{\kappa} \times \mathbf{b}) \times \hat{\mathbf{z}} - 2\hat{\mathbf{z}} \times \mathbf{u} - q\hat{\mathbf{y}}u_x, \tag{9.25}$$

where we have introduced a dimensionless wavenumber $\kappa \equiv kv_A/\Omega$ and a dimensionless frequency $\psi \equiv \omega/\Omega$. The induction equation becomes (assuming $\nabla \cdot \mathbf{u} = 0$)

$$\partial_t(B\mathbf{b}) = \nabla \times (\mathbf{u} \times \mathbf{B} + \mathbf{U} \times (B\mathbf{b})), \tag{9.26}$$

$$\partial_t\mathbf{b} = \nabla \times (\mathbf{u} \times \hat{\mathbf{z}} + U\hat{\mathbf{y}} \times \mathbf{b}), \tag{9.27}$$

$$\partial_t\mathbf{b} = \partial_z\mathbf{u} + q\Omega\hat{\mathbf{y}}b_x, \tag{9.28}$$

$$-i\psi\mathbf{b} = i\kappa_z\mathbf{u}/v_A + q\hat{\mathbf{y}}b_x, \tag{9.29}$$

$$i\psi v_A\mathbf{b} = -\kappa_z(i\mathbf{u} - q\hat{\mathbf{y}}u_x/\psi), \tag{9.30}$$

where the last line was obtained by substituting for $b_x$ back into the $y$-component of the equation. This can be substituted into the momentum equation (multiplied by $\psi$) to give:

$$-i\psi^2\mathbf{u} = -i\kappa\psi\frac{\delta P}{\rho v_A} - \kappa_z[\boldsymbol{\kappa} \times (i\mathbf{u} - q\hat{\mathbf{y}}u_x/\psi)] \times \hat{\mathbf{z}} - \psi(2\hat{\mathbf{z}} \times \mathbf{u} + q\hat{\mathbf{y}}u_x). \tag{9.31}$$

Along with the incompressibility condition we have four equations and four quantities to be eliminated, $\delta P$ and the three components of $\mathbf{u}$. Therefore:

$$-i\psi^2u_x = -i\kappa_x\psi\frac{\delta P}{\rho v_A} - i\kappa_z(\kappa_zu_x - \kappa_xu_z) + 2\psi u_y, \tag{9.32}$$

$$-i\psi^2u_y = -\kappa_z^2(iu_y - qu_x/\psi) - (2 + q)\psi u_x. \tag{9.33}$$

$$-i\psi^2u_z = -i\kappa_z\psi\frac{\delta P}{\rho v_A}. \tag{9.34}$$

From the third of these, we see that $-i\kappa_x\psi\delta P/\rho v_A = -i\psi^2u_z\kappa_x/\kappa_z$ which, using the incompressibility condition $\kappa_xu_x + \kappa_zu_z = 0$, replaces the first term on the right-hand side of the $x$-component equation and reduces the set to two equations and two variables $u_x$ and $u_y$:

$$-i\psi^2u_x = i\psi^2u_x\frac{\kappa_x^2}{\kappa_z^2} - iu_x(\kappa_z^2 + \kappa_x^2) + 2\psi u_y, \tag{9.35}$$

$$-i\psi^2u_y = -\kappa_z^2(iu_y - qu_x/\psi) - (2 + q)\psi u_x. \tag{9.36}$$

Collecting terms with $u_x$ and $u_y$ gives

$$\left(i\psi^2\frac{\kappa^2}{\kappa_z^2} - i\kappa^2\right)u_x + 2\psi u_y = 0, \tag{9.37}$$

$$\left(\frac{q\kappa_z^2}{\psi} - (2+q)\psi\right)u_x + \left(i\psi^2 - i\kappa_z^2\right)u_y = 0, \tag{9.38}$$

where $\kappa^2 = \kappa_x^2 + \kappa_z^2$. Taking the determinant of $\mathbf{A}$ in $\mathbf{A} \cdot \mathbf{u} = \mathbf{0}$ to be zero we have the following quadratic in $\psi^2$:

$$\left(\psi^2\frac{\kappa^2}{\kappa_z^2} - \kappa^2\right)\left(\psi^2 - \kappa_z^2\right) + 2\psi\left(\frac{q\kappa_z^2}{\psi} - (2+q)\psi\right) = 0, \tag{9.39}$$

which we rearrange to:

$$\frac{\kappa^2}{\kappa_z^2}\psi^4 - 2(\kappa^2 + 2 + q)\psi^2 + \kappa_z^2(\kappa^2 + 2q) = 0. \tag{9.40}$$

Solving the quadratic in $\psi^2$ we have

$$\psi^2 = \frac{\kappa_z^2}{\kappa^2}\left[\kappa^2 + 2 + q \pm \sqrt{4\kappa^2 + (2+q)^2}\right]. \tag{9.41}$$

It is straightforward to show from this that the stability condition, i.e. the condition that both roots are postive, is $2q + \kappa^2 > 0$ which is identical to (9.21) derived above. For stability at all wavenumbers we require $q > 0$, i.e. an angular velocity increasing with radius. For $q < 0$, there is instability for a range of wavenumbers $0 < \kappa^2 < 2|q|$ and the maximum growth rate is $|\psi_{\text{max}}| = |q|/2$ at a wavenumber $\kappa_{z,\text{max}}^2 = (1 + q/4)|q|$ (and $\kappa_x = 0$).

### 9.3.3. MRI: remarks

We saw above that an accretion disc with a weak vertical magnetic field suffers an instability with some minimum wavelength. From vertical force balance, we see that $H \sim \varpi c_s/v_{\text{Kep}}$ where $H$ is the thickness of the disc, $c_s$ is the sound speed and $v_{\text{Kep}}$ is the Keplerian orbit speed. Evidently, if this instability is to be effective the minimum wavelength must be less than $H$, so that since $q = -3/2$ we see from (9.21) that

$$\frac{2\pi v_A}{\sqrt{3}\,\Omega} \lesssim \frac{\varpi c_s}{v_{\text{Kep}}} \implies \beta \gtrsim \frac{4\pi^2}{3\gamma}. \tag{9.42}$$

The magnetic field must not therefore become too strong relative to the thermal pressure. Also, a strong magnetic field would be buoyantly unstable.

We looked here at the axisymmetric modes where the unperturbed field is parallel to the rotation axis; in reality we would expect the field to be dominated by its azimuthal component since this is the direction of the shear. The instability does operate on a purely azimuthal field but the modes are non-axisymmetric and the dispersion relation is more complex. Numerical non-linear analysis of this instability shows a steady-state dynamo effect as well as the desired angular momentum transport with a Shakura–Sunyaev $\alpha$ parameter of roughly the right magnitude, 0.01

to 0.1. However, many issues remain and the instability and how it works in discs is still far from fully understood.

Note that in the non-magnetic limit $v_A \to 0$ the dispersion relation becomes

$$\frac{\omega^2}{\Omega^2} = 2(2 + q)\frac{k_z^2}{k^2}, \tag{9.43}$$

so that we recover the Rayleigh stability condition $q > -2$ derived at the beginning of this section.

*And finally*

These two chapters have been a relatively short introduction to the subject. The interested reader is referred to Roberts (1967) and Spruit (2013) for more MHD!

# Exercises

9.1 **Equilibria in non-convective stars**

A non-rotating upper main-sequence star is radiative apart from a small convective core, which we can ignore here (see Braithwaite and Spruit 2017 for a review).

(a) Use the Spitzer conductivity to make an order-of-magnitude estimate of the diffusive timescale on which any magnetic field present will evolve, and compare this to the main-sequence lifetime of the star and to the dynamic (Alfvén) timescale for a magnetic field of 1 kG.

(b) The Ohmic timescale in the interior is much greater than the main-sequence lifetime of the star (as first realised by Cowling 1945), but the Ohmic timescale in the exterior is very short. Assuming that the kinetic viscosity is large, show that after some time any magnetic field present settles into an equilibrium (that is, in an energy minimum) and argue that in equilibrium, the field is potential outside the star and non-force free in the interior. (Hints: note that the plasma-$\beta$ is high inside and low outside the star, and use the vanishing force-free field theorem. Once formed, this equilibrium evolves quasi-statically on the interior Ohmic timescale which is much longer than other timescales of interest.)

(c) The equilibrium is axisymmetric. Using cylindrical coordinates ($\varpi$, $\phi$, $z$) show that the field can be expressed as the sum of poloidal and toroidal components as

$$\varpi \mathbf{B} = \nabla\psi \times \hat{\phi} + F\hat{\phi}, \tag{9.44}$$

where $\hat{\phi}$ is the azimuthal unit vector and $\psi$ and $F$ are functions of $\varpi$ and $z$, and that the poloidal and toroidal fields are associated with toroidal and poloidal currents, respectively. By considering azimuthal force balance, show that contours of $F$ in the meridional plane are parallel to those of $\psi$, i.e. that $F = F(\psi)$. (Hint: show that $(\nabla\psi) \times (\nabla F) = \mathbf{0}$.)

(d) The interiors of stars have $\beta \gg 1$. Show that an arbitrary axisymmetric magnetic field configuration can be added to a star and an equilibrium constructed by making small adjustments to the pressure and density fields. Ignore surface effects as well as any changes in the gravitational potential. Argue that this result can be generalised to non-axisymmetric fields. Note: an equilibrium does not necessarily need to be stable; it can also be unstable.

## 9.2 Kink instability in solar coronal loops

A coronal loop of magnetic field links two sunspots of opposite polarity. According to one theory of solar flares, reconnection events are triggered when a loop crosses the kink instability threshold (see e.g. Hood and Priest 1979).

(a) In a straight flux tube, the growth rate of the kink $m = 1$ instability is comparable to the Alfvén frequency, defined as $\omega_A \equiv v_A^\phi / \varpi$ where $v_A^\phi$ is the Alfvén speed associated with the azimuthal component $B_\phi$ of the magnetic field and $\varpi$ is the cylindrical radius. An axial field component $B_z$ can stabilise the field by providing an extra tension against which the instability must do work. In a tube where $B_\phi \propto \varpi$, by consideration of the force balance perpendicular to the tube axis show that the stability criterion is

$$k_z B_z > \frac{B_\phi}{\varpi}, \qquad (9.45)$$

where $k_z = 2\pi / \lambda_z$ is the wavenumber of the instability. (Hint: first show what force is required to produce the growth rate $\omega_A$ and then equate this to the restoring Lorentz force from the axial field.)

(b) We can make the approximation that the stability criterion for a curved flux tube does not differ enormously from that in a straight tube. The field between the two sunspots is initially untwisted, i.e. $B_\phi = 0$ and then one of the spots slowly rotates. Calculate the energy required to twist one sunspot up to the instability threshold by consideration of the Lorentz force in a shallow layer in the sunspot where $B_\phi$ changes from to its value in the coronal loop. As the tube is twisted, it passes quasi-statically through a series of force-free equilibria; show that the $\alpha$ parameter in the force-free equation $\nabla \times \mathbf{B} = \alpha \mathbf{B}$ increases from 0 up to some value. When the instability threshold is passed, the field in the corona relaxes back to the lowest energy state, i.e. the curl-free field $\alpha = 0$; make an estimate of the energy released in the flare and equate this to the sunspot-rotating energy calculated earlier.

# References

Balbus S A and Hawley J F 1991 A powerful local shear instability in weakly magnetized disks I — linear analysis. II — nonlinear evolution *ApJ* **376** 214–33

Blandford R D and Payne D G 1982 Hydromagnetic flows from accretion discs and the production of radio jets *Mon. Not. R. Astron. Soc.* **199** 883–903

Braithwaite J and Spruit C H 2017 Magnetic fields in non-convective regions of stars *R. Soc. Open Sci.* **4** 160271

Cassak P A, Liu Y-H and Shay M A 2017 A review of the 0.1 reconnection rate problem arXiv:1708.03449 [physics.plasm-ph]

Chandrasekhar S 1961 *Hydrodynamic and Hydromagnetic Stability* (Oxford: Clarendon)

Cowling T G 1945 On the sun's general magnetic field *Mon. Not. R. Astron. Soc.* **105** 166

de Moortel I and Browning P 2015 Recent advances in coronal heating *Phil. Trans. R. Soc. Lon.* A **373** 20140269

Hansteen V, Guerreiro N, de Pontieu B and Carlsson M 2015 Numerical simulations of coronal heating through footpoint braiding *ApJ* **811** 106

Hawley J F, Fendt C, Hardcastle M, Nokhrina E and Tchekhovskoy A 2015 Disks and jets: gravity, rotation and magnetic fields *Space Sci. Rev.* **191** 441–69

Hood A W and Priest E R 1979 Kink instability of solar coronal loops as the cause of solar flares *Solar Phys.* **64** 303–21

Roberts P H 1967 *An Introduction to Magnetohydrodynamics* (London: Longmans)

Shakura N I and Sunyaev R A 1973 Black holes in binary systems: observational appearance *Astron. Astrophys.* **24** 337–55

Spruit H C 2013 Essential magnetohydrodynamics for astrophysics arXiv: 1301.5572

Spruit H C, Foglizzo T and Stehle R 1997 Collimation of magnetically driven jets from accretion discs *Mon. Not. R. Astron. Soc.* **288** 333–42

Tayler R J 1957 Hydromagnetic instabilities of an ideally conducting fluid *Proc. Phys. Soc.* B **70** 31–48

Tomczyk S, McIntosh S W, Keil S L, Judge P G, Schad T, Seeley D H and Edmondson J 2007 Alfven waves in the solar corona *Science* **317** 1192

Uzdensky D 2014 A review of astrophysical reconnection *XL COSPAR Scientific Assembly* vol. 40 of COSPAR Meeting

van Ballegooijen A A, Asgari-Targhi M, Cranmer S R and DeLuca E E 2011 Heating of the solar chromosphere and corona by alfven wave turbulence *ApJ* **736** 3

Essential Fluid Dynamics for Scientists

**Jonathan Braithwaite**

# Appendix A

## Useful information

## A.1 Physical constants

| | | |
|---|---|---|
| speed of light | $c$ | $3 \times 10^{10}$ cm s$^{-1}$ |
| gravitational constant | $G$ | $2/3 \times 10^{-7}$ cm$^3$ g$^{-1}$ s$^{-2}$ |
| Planck constant | $h$ | $2/3 \times 10^{-26}$ erg s |
| | $\hbar \equiv h/2\pi$ | $10^{-27}$ erg s |
| Boltzmann constant | $k_B$ | $1.4 \times 10^{-16}$ erg K$^{-1}$ |
| Avogadro's number | $N_A$ | $6 \times 10^{23}$ mol$^{-1}$ |
| gas constant | $R = k_B N_A$ | $8.31 \times 10^7$ erg mol$^{-1}$ K$^{-1}$ |
| Stefan–Boltzmann constant | $\sigma_{SB} = \pi^2 k_B^4/60\hbar^3 c^2$ | $5.67 \times 10^{-5}$ erg cm$^{-2}$ s$^{-1}$ K$^{-4}$ |
| radiation constant | $a = 4\sigma/c$ | $7.6 \times 10^{-15}$ erg cm$^{-3}$ K$^{-4}$ |
| fine structure constant | $\alpha = e^2/\hbar c$ | $1/137$ |
| electron charge | $e$ | $4.8 \times 10^{-10}$ esu |
| | $e^2$ | $1.44 \times 10^{-7}$ eV cm |
| electron volt | eV | $1.6 \times 10^{-12}$ erg |
| electron mass | $m_e$ | $9 \times 10^{-28}$ g |
| | | 511 keV |
| proton mass | $m_p \approx 1$ g mol $^{-1}/N_A$ | $5/3 \times 10^{-24}$ g 938 MeV |
| proton/electron mass ratio | $m_p/m_e$ | 1836 |
| proton-neutron mass difference | $m_n - m_p$ | 1.3 MeV |
| Rydberg constant | $R_\infty = \alpha m_e c/2h$ | $1.1 \times 10^5$ cm$^{-1}$ 13.6 eV |
| Bohr radius | $a_0 = \hbar^2/m_e e^2$ | $5.3 \times 10^{-9}$ cm |
| classical electron radius | $r_0 = e^2/m_e c^2$ | $2.8 \times 10^{-13}$ cm |
| Thompson cross-section | $\sigma_T = (8\pi/3)r_0^2$ | $2/3 \times 10^{-24}$ cm$^2$ |
| Compton electron wavelength | $h/m_e c$ | $2.4 \times 10^{-10}$ cm |
| | $\hbar/m_e c$ | $3.9 \times 10^{-11}$ cm |

doi:10.1088/978-1-6817-4597-8ch10

## A.2 Astrophysical constants and units

| | | |
|---|---|---|
| Solar luminosity | $L_\odot$ | $4 \times 10^{33}$ erg s$^{-1}$ |
| Solar mass | $M_\odot$ | $2 \times 10^{33}$ g |
| Solar radius | $R_\odot$ | $7 \times 10^{10}$ cm |
| Jupiter mass | $M_J$ | $10^{-3} M_\odot$ |
| Jupiter radius | $R_J$ | $0.1 R_\odot$ |
| Earth mass | $M_\oplus$ | $3 \times 10^{-6} M_\odot$ |
| Earth radius | $R_\oplus$ | $0.009 R_\odot$ |
| astronomical unit | AU | $1.5 \times 10^{13}$ cm |
| parsec | pc | $3 \times 10^{18}$ cm |
| light year | ly | $10^{18}$ cm |
| Hubble constant | $H_0$ | $\approx 71$ km s$^{-1}$ Mpc$^{-1}$ |
| Eddington luminosity | $L_{Edd} = 4\pi c G M m / \sigma_T$ | $3.3 \times 10^4 \, (M/M_\odot) \, L_\odot$ |
| Schwarzschild radius | $r_S = 2GM/c^2$ | $3 \, (M/M_\odot)$ km |

## A.3 Vector identities and vector calculus identities

$$\mathbf{a} \cdot \mathbf{b} \times \mathbf{c} = \mathbf{c} \cdot \mathbf{a} \times \mathbf{b} = \mathbf{b} \cdot \mathbf{c} \times \mathbf{a} \tag{A.1}$$

$$\mathbf{a} \times (\mathbf{b} \times \mathbf{c}) = (\mathbf{a} \cdot \mathbf{c})\mathbf{b} - (\mathbf{a} \cdot \mathbf{b})\mathbf{c} \tag{A.2}$$

$$\nabla \times \nabla \phi = 0 \tag{A.3}$$

$$\nabla \cdot \nabla \times \mathbf{a} = 0 \tag{A.4}$$

$$\nabla^2 = \nabla \cdot \nabla, \text{ i.e. } \nabla^2 \phi = \nabla \cdot \nabla \phi \text{ and } \nabla^2 \mathbf{a} = (\nabla^2 a_x, \nabla^2 a_y, \nabla^2 a_z) \tag{A.5}$$

$$\nabla \times \nabla \times \mathbf{a} = \nabla(\nabla \cdot \mathbf{a}) - \nabla^2 \mathbf{a} \tag{A.6}$$

$$\nabla(\mathbf{a} \cdot \mathbf{b}) = (\mathbf{a} \cdot \nabla)\mathbf{b} + (\mathbf{b} \cdot \nabla)\mathbf{a} + \mathbf{a} \times \nabla \times \mathbf{b} + \mathbf{b} \times \nabla \times \mathbf{a} \tag{A.7}$$

$$\frac{1}{2}\nabla a^2 = (\mathbf{a} \cdot \nabla)\mathbf{a} + \mathbf{a} \times \nabla \times \mathbf{a} \tag{A.8}$$

$$\nabla \cdot (\mathbf{a} \times \mathbf{b}) = \mathbf{b} \cdot \nabla \times \mathbf{a} - \mathbf{a} \cdot \nabla \times \mathbf{b} \tag{A.9}$$

$$\nabla \times (\mathbf{a} \times \mathbf{b}) = \mathbf{a}(\nabla \cdot \mathbf{b}) - \mathbf{b}(\nabla \cdot \mathbf{a}) + (\mathbf{b} \cdot \nabla)\mathbf{a} - (\mathbf{a} \cdot \nabla)\mathbf{b} \tag{A.10}$$

$$\nabla \cdot (\phi \mathbf{a}) = \mathbf{a} \cdot \nabla \phi + \phi \nabla \cdot \mathbf{a} \tag{A.11}$$

$$\nabla \times (\phi \mathbf{a}) = \phi \nabla \times \mathbf{a} + \mathbf{a} \times \nabla \phi \qquad (A.12)$$

$$\nabla(\psi \phi) = \psi \nabla \phi + \phi \nabla \psi \qquad (A.13)$$

$$\oint \mathbf{a} \cdot d\mathbf{s} = \int \nabla \times \mathbf{a} \cdot d\mathbf{S} \quad \text{(Stokes' theorem)} \qquad (A.14)$$

$$\oint \mathbf{a} \cdot d\mathbf{S} = \int \nabla \cdot \mathbf{a} \, dV \quad \text{(Gauss' theorem)} \qquad (A.15)$$

## A.4 Symbols used in this book

| Quantity | Symbol | Unit |
|---|---|---|
| Time | $t$ | s |
| Pressure | $P$ | erg cm$^{-3}$ |
| Density | $\rho$ | g cm$^{-3}$ |
| Temperature | $T$ | K |
| Fluid velocity | $\mathbf{u}$ | cm s$^{-1}$ |
| components thereof | $u, v, w$ | |
| Velocity potential | $\phi$ | cm$^2$ s$^{-1}$ |
| Depth of water | $h$ | cm |
| Perturbation to surface of water | $\zeta$ | cm |
| Sound speed | $c$ (or $c_s$) | cm s$^{-1}$ |
| Specific heat at constant volume | $c_v$ | erg g$^{-1}$ K$^{-1}$ |
| Specific heat at constant pressure | $c_p$ | erg g$^{-1}$ K$^{-1}$ |
| Ratio of specific heats (a.k.a. adiabatic index) | $\gamma = c_v/c_v$ | |
| Coefficient of thermal expansivity | $\alpha$ | K$^{-1}$ |
| Coefficient of isothermal compressibility | $\kappa$ | erg$^{-1}$ cm$^3$ |
| Molar mass | $\mu_m$ | g mol$^{-1}$ |
| Specific gas constant | $R_\mu \equiv R/\mu_m$ | erg g$^{-1}$ K$^{-1}$ |
| Specific entropy | $s$ | erg g$^{-1}$ K$^{-1}$ |
| Specific internal energy | $\varepsilon$ | erg g$^{-1}$ |
| Specific enthalpy | $h = \varepsilon + p/\rho$ | erg g$^{-1}$ |
| Specific total energy | $\varepsilon = \varepsilon + u^2/2 + \Phi$ | erg g$^{-1}$ |
| Gravitational potential | $\Phi$ | erg g$^{-1}$ |
| Gravitational field | $\mathbf{g}$ | cm s$^{-2}$ |
| Magnetic field | $\mathbf{B}$ | gauss = erg$^{1/2}$ cm$^{-3/2}$ |
| Electric field | $\mathbf{E}$ | statvolt cm$^{-1}$ = erg$^{1/2}$ cm$^{-3/2}$ |

(*Continued*)

A-3

| Dynamic viscosity | $\mu$ | g cm$^{-1}$ s$^{-1}$ |
| Kinematic viscosity | $\nu$ | cm$^2$ s$^{-1}$ |
| Thermal diffusivity | $\chi$ | cm$^2$ s$^{-1}$ |
| Magnetic diffusivity | $\eta$ | cm$^2$ s$^{-1}$ |
| Charge density | $\rho_e$ | esu cm$^{-3}$ = erg $^{1/2}$ cm $^{-5/2}$ |
| Current density | **J** | esu cm$^{-2}$ s$^{-1}$ = erg $^{1/2}$ cm $^{-3/2}$ s$^{-1}$ |
| Coordinates: (spherical) radius | $r$ | |
| cylindrical radius | $\varpi$ | |
| colatitude (0 at north pole, $\pi$ at south pole) | $\theta$ | |
| azimuthal angle | $\phi$ | |

## A.5  A collection of useful results from thermodynamics

I list here a series of useful relations and concepts without going into the detail of their origin. First of all, the zeroth law is

*There are three bodies A, B and C. If A and B are both separately in equilibrium with C, then A and B are in equilibrium with each other.*

From this we can define some property of a body, temperature. If two bodies have the same temperature then they are in thermal equilibrium with each other. Furthermore, we know from experience that the state of a given mass of fluid can be completely specified by a number of parameters and that all other parameters can be worked out from the equation of state. For instance, the state of a given mass of air is completely specified by its pressure $P$ and its volume $V$. Note that pressure is an *intensive* variable as it can be measured at a particular point in space and does not depend on the size of the system, while volume is an *extensive* variable which obviously does depend on size. We could alternatively specify the state of air by pressure and specific volume $v = 1/\rho$ and then calculate other variables such as temperature from the equation of state. In the case of an ideal gas, the equation of state is

$$P = \frac{\rho R T}{\mu_m} \tag{A.16}$$

where $R$ is the universal gas constant with units erg mol$^{-1}$ K$^{-1}$. The molecular weight may vary within the gas, in which case we have three thermodynamic degrees of freedom, but in many situations (including every example in this book) it is constant and we have only two degrees of freedom, in which case it is common to write $R/\mu_m$ as $R_\mu$ so the equation of state can be written $P = \rho R_\mu T$. And in some cases we might need only one independent variable, in which case we speak of a *barotropic equation of state $\rho = \rho(P)$.*

The first law of thermodynamics is:

*If the state of an otherwise isolated system is changed by the performance of work, the amount of work needed depends solely on the change accomplished, and not on the means by which the work is performed, nor on the intermediate stages through which the system passes between its initial and final states.*

which can alternatively be expressed in the simpler form

*Energy is conserved if heat is taken into account.*

From this it is possible to demonstrate the existence of a quantity called the internal energy $U$ which is a function of state, i.e. it can be expressed as a function of the variables which describe the state of the system, for instance $U = U(P, V)$, and that changes in $U$ are given by

$$dU = dQ + dW, \tag{A.17}$$

where $dQ$ and $dW$ are the heat added to and the work done on the system.
   The second law states that

*It is impossible to devise an engine which, working in a cycle, shall produce no effect other than the transfer of heat from a colder to a hotter body.*

But note that there are several popular ways of expressing this law and that their equivalence is not always obvious at first glance. In a fluid where $P$ and $V$ are the only two independent variables

$$dU = T\,dS - P\,dV, \tag{A.18}$$

where entropy $S$ is a new function of state about which one can make the following statement:

*The entropy of an isolated system can never diminish.*

There are more terms on the right-hand side of (A.18) for systems with extra degrees of freedom; a favourite of textbooks is magnetisable systems where $-P\,dV$ is replaced or joined by $\mathbf{H} \cdot d\mathbf{M}$. In addition, it is often useful to use a new function of state—*enthalpy*, defined thus:

$$H \equiv U + PV \qquad \text{so that} \qquad dH = T\,dS + V\,dP, \tag{A.19}$$

whereby we know that it must be a function of state because it is defined as a function only of other functions of state.
   For *reversible* changes, it is possible to equate the terms of (A.17) and (A.18) and write

$$dQ = T\,dS \quad \text{and} \quad dW = -P\,dV. \tag{A.20}$$

The heat capacities are defined as $dQ/dT$ in a reversible change under various conditions. From (A.18) and (A.20) we have

$$dQ = dU + P\,dV, \tag{A.21}$$

from which we see that the heat capacites at constant volume $C_v$ and constant pressure $C_p$ are

$$C_v = T\left(\frac{\partial S}{\partial T}\right)_V = \left(\frac{\partial U}{\partial T}\right)_V \tag{A.22}$$

$$C_p = T\left(\frac{\partial S}{\partial T}\right)_P = \left(\frac{\partial U}{\partial T}\right)_P + P\left(\frac{\partial V}{\partial T}\right)_P, \tag{A.23}$$

where the forms with entropy are obtained directly from (A.20). It is fairly straightforward to prove that $C_p \geqslant C_v$, or in other words, that the ratio of the two $\gamma \equiv C_p/C_v \geqslant 1$. The heat capacities represent two properties of a fluid; we need however four properties to fully represent a fluid with two thermodynamic degrees of freedom and to predict how its state changes as the thermodynamic state changes and heat is added. The other two normally given are a measure of how the fluid expands as the temperature increases at constant pressure, and how the fluid is compressed under increasing pressure at constant temperature:

$$\alpha = \frac{1}{V}\left(\frac{\partial V}{\partial T}\right)_P \quad \text{and} \quad \kappa = -\frac{1}{V}\left(\frac{\partial V}{\partial P}\right)_T, \tag{A.24}$$

where the minus sign is added so that $\kappa$ is positive. Indeed, $\kappa$ is *always* positive; on the other hand $\alpha$ is merely almost always positive: the best known exception is water between 0 and 4 °C. From the symmetry of the way that $P$, $T$, $V$ and $S$ are related we can derive

$$-\left(\frac{\partial T}{\partial V}\right)_S = \left(\frac{\partial P}{\partial S}\right)_V \quad \text{and} \quad \left(\frac{\partial T}{\partial P}\right)_S = \left(\frac{\partial V}{\partial S}\right)_P,$$

$$\left(\frac{\partial S}{\partial V}\right)_T = \left(\frac{\partial P}{\partial T}\right)_V \quad \text{and} \quad -\left(\frac{\partial S}{\partial P}\right)_T = \left(\frac{\partial V}{\partial T}\right)_P, \tag{A.25}$$

which are normally referred to as Maxwell's relations. These relations together with two results from mathematics

$$\left(\frac{\partial x}{\partial y}\right)_z\left(\frac{\partial y}{\partial x}\right)_z = 1 \quad \text{and} \quad \left(\frac{\partial x}{\partial y}\right)_z\left(\frac{\partial y}{\partial z}\right)_x\left(\frac{\partial z}{\partial x}\right)_y = -1 \tag{A.26}$$

we can use to convert between any partial derivatives of $P$, $T$, $V$ and $S$ and express them in terms of $C_v$, $C_p$, $\alpha$ and $\kappa$.

Extensive quantities such as $V$ and $C_p$ can be made intensive by dividing them by the mass of the system, giving in these two examples the *specific volume* $v$ (equal of course to $1/\rho$) and *specific heat capacity* $c_p$.

In general $c_v$, $c_p$, $\alpha$ and $\kappa$ are functions of the thermodynamic state of a fluid but in ideal gas they are constant:

$$\text{(Ideal gas)} \quad C_v = \frac{nR_\mu}{2}, \quad c_p = \frac{(n+2)R_\mu}{2}, \qquad \alpha = \frac{1}{T} \quad \text{and} \quad \kappa = \frac{1}{P}, \text{ (A.27)}$$

where $n$ is the degrees of freedom, equal to 3 in a monatomic gas and 5 in a diatomic gas. In addition, since $\kappa = 1$ we have the very useful result

$$\text{(Ideal gas)} \quad PV^\gamma = \text{const} \quad \text{or alternatively} \quad P \propto \rho^\gamma \qquad \text{(A.28)}$$

for adiabatic changes.

www.ingramcontent.com/pod-product-compliance
Lightning Source LLC
Chambersburg PA
CBHW081538220326
41598CB00036B/6481